WWS 320

Prof Lee Silver

# OUR POSTHUMAN FUTURE

# OUR POSTHUMAN FUTURE

·

CONSEQUENCES

OF THE

BIOTECHNOLOGY REVOLUTION

·

FRANCIS FUKUYAMA

FARRAR, STRAUS AND GIROUX

NEW YORK

Farrar, Straus and Giroux
19 Union Square West, New York 10003

Copyright © 2002 by Francis Fukuyama
All rights reserved
Distributed in Canada by Douglas & McIntyre Ltd.
Printed in the United States of America
First edition, 2002

ISBN: 0-374-23643-7
Library of Congress Control Number: 2002100914

Designed by Abby Kagan

www.fsgbooks.com

1 3 5 7 9 10 8 6 4 2

**For John Sebastian, last but not least**

Enough: the time is coming when politics will have a different meaning.
Friedrich Nietzsche, *The Will to Power*, Section 960[1]

# CONTENTS

W riting a book on biotechnology might seem to be quite a leap for someone who in recent years has been interested primarily in issues of culture and economics, but there is actually a method to this madness.

In early 1999, I was asked by Owen Harries, editor of *The National Interest*, to write a ten-year retrospective on my article "The End of History?" which he had originally published in the summer of 1989. In that article I argued that Hegel had been right in saying that history had ended in 1806, since there had been no essential political progress beyond the principles of the French Revolution, which he had seen consolidated by Napoleon's victory in the Battle of Jena that year. The collapse of communism in 1989 signaled only the denouement of a broader convergence toward liberal democracy around the globe.

In the course of thinking through the many critiques of that original piece that had been put forward, it seemed to me that the only one that was not possible to refute was the argument that there could be no end of history unless there was an end of science. As I had described the mechanism of a progressive universal history in my subsequent book *The End of History and the Last Man*, the unfolding of modern natural science and the technology that it spawns emerges as one of its chief drivers. Much of late-twentieth-century technology, like the so-called Information Revolution, was quite conducive to the spread of liberal democracy. But we are nowhere near the end of science, and indeed seem to be in the midst of a monumental period of advance in the life sciences.

I had in any event been thinking about the impact of modern biology on our understanding of politics for some time. This sprang out of a study group that I ran for several years on the impact of new sciences on international politics. Some of my initial thinking on this issue was reflected in my book *The Great Disruption*, which dealt with the question of human nature and norms, and how our understanding of them was shaped by new empirical information from fields like ethology, evolutionary biology, and cognitive neuroscience. But the invitation to write a retrospective on the "end of history" was the occasion to begin thinking about the future in a more systematic way, which resulted in an article published in *The National Interest* in 1999 entitled "Second Thoughts: The Last Man in a Bottle." The present volume is a vast expansion of the themes first undertaken there.

The terrorist attacks on the United States of September 11, 2001, again raised doubts about the end-of-history thesis, this time on the grounds that we were witnessing a "clash of civilizations" (to use Samuel P. Huntington's phrase) between the West and Islam. I believe that these events prove nothing of the sort, and that the Islamic radicalism driving these attacks is a desperate rearguard action that will in time be overwhelmed by the broader tide of modernization. What these events point to, however, is the fact that science and technology, from which the modern world springs, themselves represent our civilization's key vulnerabilities. Airliners, skyscrapers, and biology labs—all symbols of modernity—were turned into weapons in a stroke of malign ingenuity. The current volume does not deal with biological weapons, but the emergence of bioterrorism as a live threat points to the need, outlined in this book, for greater political control over the uses of science and technology.

There are, needless to say, many people who helped me on this project whom I would like to thank. These include David Armor, Larry Arnhart, Scott Barrett, Peter Berkowitz, Mary Cannon, Steve Clemons, Eric Cohen, Mark Cordover, Richard Doerflinger, Bill Drake, Terry Eastland, Robin Fox, Hillel Fradkin, Andrew Franklin, Franco Furger, Jonathan Galassi, Tony Gilland, Richard Hassing, Richard Hayes, George Holmgren, Leon Kass, Bill Kristol, Jay Lefkowitz, Mark Lilla, Michael Lind, Michael McGuire, David Prentice, Gary Schmitt, Abram Shulsky, Gregory Stock, Richard Velkley, Caroline Wagner, Marc Wheat, Edward O. Wilson, Adam Wolfson, and Robert Wright. I am very grateful to my literary agent, Esther Newberg, and to all of those at International Creative Management who have helped me over the years. My research assistants, Mike Curtis, Ben Allen, Christine Pommerening, Sanjay Marwah, and Brian Grow, provided invaluable assistance. I would like to thank the Bradley Foundation for providing support for student fellowships as part of this project. Cynthia Paddock, my all-around assistant, contributed to the final production of the manuscript. As always, my wife, Laura, was a thoughtful commentator on the manuscript, on issues where she has very strong views.

**PART I**

·

PATHWAYS TO THE

FUTURE

# A TALE OF TWO DYSTOPIAS

**The threat to man does not come in the first instance from the potentially lethal machines and apparatus of technology. The actual threat has always afflicted man in his essence. The rule of enframing (*Gestell*) threatens man with the possibility that it could be denied to him to enter into a more original revealing and hence to experience the call of a more primal truth.**

**Martin Heidegger, *The Question Concerning Technology*[1]**

I was born in 1952, right in the middle of the American baby boom. For any person growing up as I did in the middle decades of the twentieth century, the future and its terrifying possibilities were defined by two books, George Orwell's *1984* (first published in 1949) and Aldous Huxley's *Brave New World* (published in 1932).

The two books were far more prescient than anyone realized at the time, because they were centered on two different technologies that would in fact emerge and shape the world over the next two generations. The novel *1984* was about what we now call information technology: central to the success of the vast, totalitarian empire that had been set up over Oceania was a device called the telescreen, a wall-sized flat-panel display that could simultaneously send and receive images from each individual household to a hovering Big Brother. The telescreen was what permitted the vast centralization of social life under the Ministry of Truth and the Ministry of Love, for it allowed the government to banish privacy by monitoring every word and deed over a massive network of wires.

*Brave New World,* by contrast, was about the other big technological revolution about to take place, that of biotechnology. Bokanovskification, the hatching of people not in wombs but, as we now say, in vitro; the drug soma, which gave people instant happiness; the Feelies, in which sensation was simulated by implanted electrodes; and the modification of behavior through constant subliminal repetition and, when that didn't work, through the administration of various artificial hormones were what gave this book its particularly creepy ambiance.

With at least a half century separating us from the publication of these books, we can see that while the technological predictions they made were startlingly accurate, the political predictions of the first book, *1984,* were entirely wrong. The year 1984 came and went, with the United States still locked in a Cold War struggle with the Soviet Union. That year saw the introduction of a new model of the IBM personal computer and the beginning of what became the PC revolution. As Peter Huber has argued, the personal computer, linked to the Internet, was in fact the realization of Orwell's telescreen.[2] But instead of becoming an instrument of centralization and tyranny, it led to just the opposite: the democratization of access to information and the decentralization of politics. Instead of Big Brother watching everyone, people could use the PC and Internet to watch Big Brother, as governments everywhere were driven to publish more information on their own activities.

Just five years after 1984, in a series of dramatic events that would earlier have seemed like political science fiction, the Soviet Union and its empire collapsed, and the totalitarian threat that Orwell had so vividly evoked vanished. People were again quick to point out that these two events—the collapse of totalitarian empires and the emergence of the personal computer, as well as other forms of inexpensive information technology, from TVs and radios to faxes and e-mail—were not unrelated. Totalitarian rule depended on a regime's ability to maintain a monopoly over information, and once modern information technology made that impossible, the regime's power was undermined.

The political prescience of the other great dystopia, Brave New World, remains to be seen. Many of the technologies that Huxley envisioned, like in vitro fertilization, surrogate motherhood, psychotropic drugs, and genetic engineering for the manufacture of children, are already here or just over the horizon. But this revolution has only just begun; the daily avalanche of announcements of new breakthroughs in biomedical technology and achievements such as the completion of the Human Genome Project in the year 2000 portend much more serious changes to come.

Of the nightmares evoked by these two books, Brave New World's always struck me as more subtle and more challenging. It is easy to see what's wrong with the world of 1984: the protagonist, Winston Smith, is known to hate rats above all things, so Big Brother devises a cage in which rats can bite at Smith's face in order to get him to betray his lover. This is the world of classical tyranny, technologically empowered but not so different from what we have tragically seen and known in human history.

In Brave New World, by contrast, the evil is not so obvious because no one is hurt; indeed, this is a world in which everyone gets what they want. As one of the characters notes, "The Controllers realized that force was no good," and that people would have to be seduced rather than compelled to live in an orderly society. In this world, disease and social conflict have been abolished, there is no depression, madness, loneliness, or emotional distress, sex is good and readily available. There is even a government ministry to ensure that the length of time between the appearance of a desire and its satisfaction is kept to a minimum. No one takes religion seriously any longer, no one is introspective or has unrequited longings, the biological family has been abolished, no one reads Shakespeare. But no one (save John the Savage, the book's protagonist) misses these things, either, since they are happy and healthy.

Since the novel's publication, there have probably been several million high school essays written in answer to the question, "What's wrong with this picture?" The answer given (on papers that get A's, at any rate) usually runs something like this: the people in *Brave New World* may be healthy and happy, but they have ceased to be *human beings.* They no longer struggle, aspire, love, feel pain, make difficult moral choices, have families, or do any of the things that we traditionally associate with being human. They no longer have the characteristics that give us human dignity. Indeed, there is no such thing as the human race any longer, since they have been bred by the Controllers into separate castes of Alphas, Betas, Epsilons, and Gammas who are as distant from each other as humans are from animals. Their world has become unnatural in the most profound sense imaginable, because *human nature* has been altered. In the words of bioethicist Leon Kass, "Unlike the man reduced by disease or slavery, the people dehumanized à la *Brave New World* are not miserable, don't know that they are dehumanized, and, what is worse, would not care if they knew. They are, indeed, happy slaves with a slavish happiness."[3]

But while this kind of answer is usually adequate to satisfy the typical high school English teacher, it does not (as Kass goes on to note) probe nearly deeply enough. For one can then ask, What is so important about being a human being in the traditional way that Huxley defines it? After all, what the human race is today is the product of an evolutionary process that has been going on for millions of years, one that with any luck will continue well into the future. There are no fixed human characteristics, except for a general capability to choose what we want to be, to modify ourselves in accordance with our desires. So who is to tell us that being human and having dignity means sticking with a set of emotional responses that are the accidental by-product of our evolutionary history? There is no such thing as a biological family, no such thing as human nature or a "normal" human being, and even if there were, why should that be a guide for what is right and just? Huxley is telling us, in effect, that we should continue to feel pain, be depressed or lonely, or suffer from debilitating disease, all because that is what human beings have done for most of their existence as a species. Certainly, no one ever got elected to Congress on such a platform. Instead of taking these characteristics and saying that they are the basis for "human dignity," why don't we simply accept our destiny as creatures who modify themselves?

Huxley suggests that one source for a definition of what it means to be a human being is religion. In *Brave New World*, religion has been abolished and Christianity is a distant memory. The Christian tradition maintains that man is created in God's image, which is the source of human dignity. To use biotechnology to engage in what another Christian writer, C. S. Lewis, called the "abolition of man" is thus a violation of God's will. But I don't think that a careful reading of Huxley or Lewis leads to the conclusion that either writer believed religion to be the *only* grounds on which one could understand the meaning of being human. Both writers suggest that nature itself, and in particular human nature, has a special role in defining for us what is right and wrong, just and unjust, important and unimportant. So our final judgment on "what's wrong" with Huxley's brave new world stands or falls with our view of how important human nature is as a source of values.

The aim of this book is to argue that Huxley was right, that the most significant threat posed by contemporary biotechnology is the possibility that it will alter human nature and thereby move us into a "posthuman" stage of history. This is important, I will argue, because human nature exists, is a meaningful concept, and has provided a stable continuity to our experience as a species. It is, conjointly with religion, what defines our most basic values. Human nature shapes and constrains the possible kinds of political regimes, so a technology powerful enough to reshape what we are will have possibly malign consequences for liberal democracy and the nature of politics itself.

It may be that, as in the case of *1984*, we will eventually find biotechnology's consequences are completely and surprisingly benign, and that we were wrong to lose sleep over it. It may be that the technology will in the end prove much less powerful than it seems today, or that people will be moderate and careful in their application of it. But one of the reasons I am not quite so sanguine is that biotechnology, in contrast to many other scientific advances, mixes obvious benefits with subtle harms in one seamless package.

Nuclear weapons and nuclear energy were perceived as dangerous from the start, and therefore were subject to strict regulation from the moment the Manhattan Project created the first atomic bomb in 1945. Observers like Bill Joy have worried about nanotechnology—that is, molecular-scale self-replicating machines capable of reproducing out of control and destroying their creators.[4] But such threats are actually the easiest to deal with because they are so obvious. If you are likely to be killed by a machine you've created, you take measures to protect yourself. And so far we've had a reasonable record in keeping our machines under control.

There may be products of biotechnology that will be similarly obvious in the dangers they pose to mankind—for example, superbugs, new viruses, or genetically modified foods that produce toxic reactions. Like nuclear weapons or nanotechnology, these are in a way the easiest to deal with because once we have identified them as dangerous, we can treat them as a straightforward threat. The more typical threats raised by biotechnology, on the other hand, are those captured so well by Huxley, and are summed up in the title of an article by novelist Tom Wolfe, "Sorry, but Your Soul Just Died."[5] Medical technology offers us in many cases a devil's bargain: longer life, but with reduced mental capacity; freedom from depression, together with freedom from creativity or spirit; therapies that blur the line between what we achieve on our own and what we achieve because of the levels of various chemicals in our brains.

Consider the following three scenarios, all of which are distinct possibilities that may unfold over the next generation or two.

The first has to do with new drugs. As a result of advances in neuropharmacology, psychologists discover that human personality is much more plastic than formerly believed. It is already the case that psychotropic drugs such as Prozac and Ritalin can affect traits like self-esteem and the ability to concentrate, but they tend to produce a host of unwanted side effects and hence are shunned except in cases of clear therapeutic need. But in the future, knowledge of genomics permits pharmaceutical companies to tailor drugs very specifically to the genetic profiles of individual patients and greatly minimize unintended side effects. Stolid people can become vivacious; introspective ones extroverted; you can adopt one personality on Wednesday and another for the weekend. There is no longer any excuse for anyone to be depressed or unhappy; even "normally" happy people can make themselves happier without worries of addiction, hangovers, or long-term brain damage.

In the second scenario, advances in stem cell research allow scientists to regenerate virtually any tissue in the body, such that life expectancies are pushed well above 100 years. If you need a new heart or liver, you just grow one inside the chest cavity of a pig or cow; brain damage from Alzheimer's and stroke can be reversed. The only problem is that there are many subtle and some not-so-subtle aspects of human aging that the biotech industry hasn't quite figured out how to fix: people grow mentally rigid and increasingly fixed in their views as they age, and try as they might, they can't make themselves sexually attractive to each other and continue to long for partners of reproductive age. Worst of all, they just refuse to get out of the way, not just of their children, but their grandchildren and great-grandchildren. On the other hand, so few people have children or any connection with traditional reproduction that it scarcely seems to matter.

In a third scenario, the wealthy routinely screen embryos before implantation so as to optimize the kind of children they have. You can increasingly tell the social background of a young person by his or her looks and intelligence; if someone doesn't live up to social expectations, he tends to blame bad genetic choices by his parents rather than himself. Human genes have been transferred to animals and even to plants, for research purposes and to produce new medical products; and animal genes have been added to certain embryos to increase their physical endurance or resistance to disease. Scientists have not dared to produce a full-scale chimera, half human and half ape, though they could; but young people begin to suspect that classmates who do much less well than they do are in fact genetically not fully human. Because, in fact, they aren't.

Sorry, but your soul just died . . .

Toward the very end of his life, Thomas Jefferson wrote, "The general spread of the light of science has already laid open to every view the palpable truth, that the mass of mankind has not been born with saddles on their backs, nor a favored few booted and spurred, ready to ride them legitimately, by the grace of God."[6] The political equality enshrined in the Declaration of Independence rests on the empirical fact of natural human equality. We vary greatly as individuals and by culture, but we share a common humanity that allows every human being to potentially communicate with and enter into a moral relationship with every other human being on the planet. The ultimate question raised by biotechnology is, What will happen to political rights once we are able to, in effect, breed some people with saddles on their backs, and others with boots and spurs?

## A STRAIGHTFORWARD SOLUTION

What should we do in response to biotechnology that in the future will mix great potential benefits with threats that are either physical and overt or spiritual and subtle? The answer is obvious: *We should use the power of the state to regulate it.* And if this proves to be beyond the power of any individual nation-state to regulate, it needs to be regulated on an international basis. We need to start thinking concretely now about how to build institutions that can discriminate between good and bad uses of biotechnology, and effectively enforce these rules both nationally and internationally.

This obvious answer is not obvious to many of the participants in the current biotechnology debate. The discussion remains mired at a relatively abstract level about the ethics of procedures like cloning or stem cell research, and divided into one camp that would like to permit everything and another camp that would like to ban wide areas of research and practice. The broader debate is of course an important one, but events are moving so rapidly that we will soon need more practical guidance on how we can direct future developments so that the technology remains man's servant rather than his master. Since it seems very unlikely that we will either permit everything or ban research that is highly promising, we need to find a middle ground.

The creation of new regulatory institutions is not something that should be undertaken lightly, given the inefficiencies that surround all efforts at regulation. For the past three decades, there has been a commendable worldwide movement to deregulate large sectors of every nation's economy, from airlines to telecommunications, and more broadly to reduce the size and scope of government. The global economy that has emerged as a result is a far more efficient generator of wealth and technological innovation. Excessive regulation in the past has led many to become instinctively hostile to state intervention in any form, and it is this knee-jerk aversion to regulation that will be one of the chief obstacles to getting human biotechnology under political control.

But it is important to discriminate: what works for one sector of the economy will not work for another. Information technology, for example, produces many social benefits and relatively few harms and therefore has appropriately gotten by with a fairly minimal degree of government regulation. Nuclear materials and toxic waste, on the other hand, are subject to strict national and international controls because unregulated trade in them would clearly be dangerous.

One of the biggest problems in making the case for regulating human biotechnology is the common view that even if it were desirable to stop technological advance, it is impossible to do so. If the United States or any other single country tries to ban human cloning or germline genetic engineering or any other procedure, people who wanted to do these things would simply move to a more favorable jurisdiction where they were permitted. Globalization and international competition in biomedical research ensure that countries that hobble themselves by putting ethical constraints on their scientific communities or biotechnology industries will be punished.

The idea that it is impossible to stop or control the advance of technology is simply wrong, for reasons that will be laid out more fully in Chapter 10 of this book. We in fact control all sorts of technologies and many types of scientific research: people are no more free to experiment in the development of new biological warfare agents than they are to experiment on human subjects without the latter's informed consent. The fact that there are some individuals or organizations that violate these rules, or that there are countries where the rules are either nonexistent or poorly enforced, is no excuse for not making the rules in the first place. People get away with robbery and murder, after all, which is not a reason to legalize theft and homicide.

We need at all costs to avoid a defeatist attitude with regard to technology that says that since we can't do anything to stop or shape developments we don't like, we shouldn't bother trying in the first place. Putting in place a regulatory system that would permit societies to control human biotechnology will not be easy: it will require legislators in countries around the world to step up to the plate and make difficult decisions on complex scientific issues. The shape and form of the institutions designed to implement new rules is a wide-open question; designing them to be minimally obstructive of positive developments while giving them effective enforcement capabilities is a significant challenge. Even more challenging will be the creation of common rules at an international level, the forging of a consensus among countries with different cultures and views on the underlying ethical questions. But political tasks of comparable complexity have been successfully undertaken in the past.

## BIOTECHNOLOGY AND THE RECOMMENCEMENT OF HISTORY

Many of the current debates over biotechnology, on issues like cloning, stem cell research, and germ-line engineering, are polarized between the scientific community and those with religious commitments. I believe that this polarization is unfortunate because it leads many to believe that the *only* reason one might object to certain advances in biotechnology is out of religious belief. Particularly in the United States, biotechnology has been drawn into the debate over abortion; many researchers feel that valuable progress is being checked out of deference to a small number of antiabortion fanatics.

I believe that it is important to be wary of certain innovations in biotechnology for reasons that have nothing to do with religion. The case that I will lay out here might be called Aristotelian, not because I am appealing to Aristotle's authority as a philosopher, but because I take his mode of rational philosophical argument about politics and nature as a model for what I hope to accomplish.

Aristotle argued, in effect, that human notions of right and wrong—what we today call human rights—were ultimately based on human nature. That is, without understanding how natural desires, purposes, traits, and behaviors fit together into a human whole, we cannot understand human ends or make judgments about right and wrong, good and bad, just and unjust. Like many more recent utilitarian philosophers, Aristotle believed that the good was defined by what people desired; but while utilitarians seek to reduce human ends to a simple common denominator like the relief of suffering or the maximization of pleasure, Aristotle retained a complex and nuanced view of the diversity and greatness of natural human ends. The purpose of his philosophy was to try to differentiate the natural from the conventional, and to rationally order human goods.

Aristotle, together with his immediate predecessors Socrates and Plato, initiated a dialogue about the nature of human nature that continued in the Western philosophical tradition right up to the early modern period, when liberal democracy was born. While there were significant disputes over what human nature was, no one contested its importance as a basis for rights and justice. Among the believers in natural right were the American Founding Fathers, who based their revolution against the British crown on it. Nonetheless, the concept has been out of favor for the past century or two among academic philosophers and intellectuals.

As we will see in Part II of this book, I believe this is a mistake, and that any meaningful definition of rights must be based on substantive judgments about human nature. Modern biology is finally giving some meaningful empirical content to the concept of human nature, just as the biotech revolution threatens to take the punch bowl away.

Whatever academic philosophers and social scientists may think of the concept of human nature, the fact that there has been a stable human nature throughout human history has had very great political consequences. As Aristotle and every serious theorist of human nature has understood, human beings are by nature cultural animals, which means that they can learn from experience and pass on that learning to their descendants through nongenetic means. Hence human nature is not narrowly determinative of human behavior but leads to a huge variance in the way people raise children, govern themselves, provide resources, and the like. Mankind's constant efforts at cultural self-modification are what lead to human history and to the progressive growth in the complexity and sophistication of human institutions over time.

The fact of progress and cultural evolution led many modern thinkers to believe that human beings were almost infinitely plastic— that is, that they could be shaped by their social environment to behave in open-ended ways. It is here that the contemporary prejudice against the concept of human nature starts. Many of those who believed in the social construction of human behavior had strong ulterior motives: they hoped to use social engineering to create societies that were just or fair according to some abstract ideological principle. Beginning with the French Revolution, the world has been convulsed with a series of utopian political movements that sought to create an earthly heaven by radically rearranging the most basic institutions of society, from the family to private property to the state. These movements crested in the twentieth century, with the socialist revolutions that took place in Russia, China, Cuba, Cambodia, and elsewhere.

By the end of the century, virtually every one of these experiments had failed, and in their place came efforts to create or restore equally modern but less politically radical liberal democracies. One important reason for this worldwide convergence on liberal democracy had to do with the tenacity of human nature. For while human behavior is plastic and variable, it is not infinitely so; at a certain point deeply rooted natural instincts and patterns of behavior reassert themselves to undermine the social engineer's best-laid plans. Many socialist regimes abolished private property, weakened the family, and demanded that people be altruistic to mankind in general rather than to a narrower circle of friends and family. But evolution did not shape human beings in this fashion. Individuals in socialist societies resisted the new institutions at every turn, and when socialism collapsed after the fall of the Berlin Wall in 1989, older, more familiar patterns of behavior reasserted themselves everywhere.

Political institutions cannot abolish either nature or nurture altogether and succeed. The history of the twentieth century was defined by two opposite horrors, the Nazi regime, which said biology was everything, and communism, which maintained that it counted for next to nothing. Liberal democracy has emerged as the only viable and legitimate political system for modern societies because it avoids either extreme, shaping politics according to historically created norms of justice while not interfering excessively with natural patterns of behavior.

There were many other factors affecting the trajectory of history, which I discussed in my book *The End of History and the Last Man*.[7] One of the basic drivers of the human historical process has been the development of science and technology, which is what determines the horizon of economic production possibilities and therefore a great deal of a society's structural characteristics. The development of technology in the late twentieth century was particularly conducive to liberal democracy. This is not because technology promotes political freedom and equality per se—it does not—but because late-twentieth-century technologies (particularly those related to information) are what political scientist Ithiel de Sola Pool has labeled technologies of freedom.[8]

There is no guarantee, however, that technology will always produce such positive political results. Many technological advances of the past reduced human freedom.[9] The development of agriculture, for example, led to the emergence of large hierarchical societies and made slavery more feasible than it had been in hunter-gatherer times. Closer to our own time, Eli Whitney's invention of the cotton gin made cotton a significant cash crop in the American South at the beginning of the nineteenth century and led to the revitalization of the institution of slavery there.

As the more perceptive critics of the concept of the "end of history" have pointed out, there can be no end of history without an end of modern natural science and technology.[10] Not only are we not at an end of science and technology; we appear to be poised at the cusp of one of the most momentous periods of technological advance in history. Biotechnology and a greater scientific understanding of the human brain promise to have extremely significant political ramifications. Together, they reopen possibilities for social engineering on which societies, with their twentieth-century technologies, had given up.

If we look back at the tools of the past century's social engineers and utopian planners, they seem unbelievably crude and unscientific. Agitprop, labor camps, reeducation, Freudianism, early childhood conditioning, behavioralism—all of these were techniques for pounding the square peg of human nature into the round hole of social planning. None of them were based on knowledge of the neurological structure or biochemical basis of the brain; none understood the genetic sources of behavior, or if they did, none could do anything to affect them.

All of this may change in the next generation or two. We do not have to posit a return of state-sponsored eugenics or widespread genetic engineering to see how this could happen. Neuropharmacology has already produced not just Prozac for depression but Ritalin to control the unruly behavior of young children. As we discover not just correlations but actual molecular pathways between genes and traits like intelligence, aggression, sexual identity, criminality, alcoholism, and the like, it will inevitably occur to people that they can make use of this knowledge for particular social ends. This will play itself out as a series of ethical questions facing individual parents, and also as a political issue that may someday come to dominate politics. If wealthy parents suddenly have open to them the opportunity to increase the intelligence of their children as well as that of all their subsequent descendants, then we have the makings not just of a moral dilemma but of a full-scale class war.

This book is divided into three parts. The first lays out some plausible pathways to the future and draws some first-order consequences, from those that are near-term and very likely through those that are more distant and uncertain. The four stages outlined here are:

- increasing knowledge about the brain and the biological sources of human behavior;
- neuropharmacology and the manipulation of emotions and behavior;
- the prolongation of life;
- and finally, genetic engineering.

Part II deals with the philosophical issues raised by an ability to manipulate human nature. It argues for the centrality of human nature to our understanding of right and wrong—that is, human rights—and how we can develop a concept of human dignity that does not depend on religious assumptions about the origins of man. Those not inclined to more theoretical discussions of politics may choose to skip over some of the chapters here.

The final part is more practical: it argues that if we are worried about some of the long-term consequences of biotechnology, we can do something about it by establishing a regulatory framework to separate legitimate and illegitimate uses. This part of the book may seem to have the opposite vice from Part II, getting into the details of specific agencies and laws in the United States and other countries, but there is a reason for this. The advance of technology is so rapid that we need to move quickly to much more concrete analysis of what kinds of institutions will be required to deal with it.

There are many near-term practical and policy-related issues that have been raised by advances in biotechnology such as the completion of the Human Genome Project, including genetic discrimination and the privacy of genetic information. This book will not focus on any of these questions, partly because they have been dealt with extensively by others, and partly because the biggest challenges opened up by biotechnology are not those immediately on the horizon but the ones that may be a decade to a generation or more away. What is important to recognize is that this challenge is not merely an ethical one but a political one as well. For it will be the political decisions that we make in the next few years concerning our relationship to this technology that determine whether or not we enter into a posthuman future and the potential moral chasm that such a future opens before us.

**2**

## SCIENCES OF THE BRAIN

❨

**W**hat are the prospects that the biotech revolution will have *political* consequences, as opposed to simply affecting the lives of individual parents and children? What new possibilities will exist for modifying or controlling human behavior on a macro level, and in particular, how likely is it that we might someday be able to consciously modify human nature?

Some promoters of the Human Genome Project, such as Human Genome Science's CEO William Haseltine, have made far-reaching claims about what contemporary molecular biology will achieve, arguing that "as we understand the body's repair process at the genetic level . . . we will be able to advance the goal of maintaining our bodies in normal function, perhaps perpetually."[1] But most scientists working in the field have much more modest views of what they are doing and what they may someday achieve. Many would assert that they are simply seeking remedies for certain genetically linked diseases like breast cancer and cystic fibrosis, that there are immense obstacles to human cloning and genetic enhancement, and that the modification of human nature is the stuff of science fiction, not technological possibility.

Technological prediction is notoriously difficult and risky, particularly when talking about events that may still lie a generation or two away. Nonetheless, it is important to lay out some scenarios for possible futures that suggest a range of outcomes, some of which are very likely and even emerging today, and others which may never in the end materialize. As we shall see, modern biotechnology has *already* produced effects that will have consequences for world politics in the coming generation, even if genetic engineering fails to produce a single designer baby before then.

In speaking about the biotech revolution, it is important to remember that we are talking about something much broader than genetic engineering. What we are living through today is not simply a technological revolution in our ability to decode and manipulate DNA, but a revolution in the underlying science of biology. This scientific revolution draws on findings and advances in a number of related fields besides molecular biology, including cognitive neuroscience, population genetics, behavior genetics, psychology, anthropology, evolutionary biology, and neuropharmacology. All of these areas of scientific advance have potential political implications, because they enhance our knowledge of, and hence our ability to manipulate, the source of all human behavior, the brain.

As we shall see, the world could look very different in the coming decades without our having to resort to heroic assumptions about the possibilities for genetic engineering. Today and in the very near future, we face ethical choices about genetic privacy, the proper uses of drugs, research involving embryos, and human cloning. Soon, however, we are going to face issues about embryo selection and the degree to which all medical technologies can be used for enhancement rather than therapeutic purposes.

## THE REVOLUTION IN COGNITIVE NEUROSCIENCE

The first pathway to the future has nothing to do with technology but simply with the accumulation of knowledge about genetics and behavior. Many of the currently anticipated benefits from the Human Genome Project are not related to potential genetic engineering but rather arise from genomics—that is, an understanding of the underlying functions of genes. Genomics will permit, for example, the tailoring of drugs to particular individuals to reduce the chances of unwanted side effects; it will give plant breeders far more precise knowledge in the design of new species.[2]

The attempt to link genes to behavior predates the Human Genome Project by many years, however, and has already resulted in a number of pitched political battles.

Since at least the time of the ancient Greeks, human beings have been arguing over the relative importance of nature versus nurture in human behavior. For much of the twentieth century, the natural and particularly the social sciences tended to emphasize the cultural drivers of behavior at the expense of natural ones. The pendulum has been swinging backward—many would argue, too far backward—in recent years, in favor of genetic causes.[3] This shift in scientific outlook is reflected everywhere in the popular press, with discussion of "genes for" everything from intelligence to fatness to aggression.

The debate over the roles of heredity and culture in the shaping of human outcomes has been a highly politicized one from the start, with conservatives tending to favor explanations based on nature, and the Left emphasizing the role of nurture. Hereditarian arguments were badly misused by various racists and bigots through the early decades of the twentieth century to explain why some races, cultures, and societies were inferior to others. Hitler is only the most famous right-wing champion of genetic thinking. Opponents of immigration to the United States before passage of the restrictive Immigration Act of 1924 argued, like Madison Grant in his 1921 book *The Passing of the Great Race*,[4] that the shift in immigration patterns from northern to southern Europe meant a deterioration of the American racial stock.[5]

The dubious pedigree of hereditarian arguments cast a pall over most discussions of genetics during the second half of the twentieth century. Progressive intellectuals were particularly intent on beating back arguments about nature. This was not only because natural differences between groups of people implied social hierarchy, but also because natural characteristics, even when universally shared, implied limits to human plasticity, and hence to human hopes and aspirations. Feminists were among the most fierce resisters of any suggestion that male-female differences were genetic rather than socially constructed.[6]

The problem with the extreme social constructionist view and the extreme hereditarian view is that neither is tenable in the light of currently available empirical evidence. In the process of mobilizing for World War I, the U.S. Army began widespread intelligence testing of new recruits, for the first time providing data on the cognitive abilities of different racial and ethnic groups.[7] These data were seized on by opponents of immigration as evidence for the mental inferiority of, among others, Jews and blacks. In one of the great early defeats of "scientific racism," the anthropologist Franz Boas showed in a carefully constructed study that immigrant children's head sizes and intelligence converged on those of the native-born when the children were fed an American diet. Others demonstrated the cultural bias embedded in the army intelligence tests (the tests asked children to identify, among other things, tennis courts, which most immigrant children had never seen).

On the other hand, any parent who has raised siblings knows from experience that there are many individual differences that simply cannot be explained in terms of upbringing and environment. Until now, there have been only two ways to scientifically disentangle natural from cultural causes of behavior. The first is through the discipline of behavior genetics, and the second through cross-cultural anthropology. The future, however, almost inevitably promises far more precise empirical knowledge of the molecular and neural pathways leading from genes to behavior.

Behavior genetics is based on the study of twins—ideally, of identical twins reared apart. (These are referred to as monozygotic twins because they come from a division of the same fertilized egg.) We know that identical twins have the same genotype—that is, the same DNA—and assume that the differences that subsequently emerge in the behavior of the twins reflect the different environments in which they are raised rather than heredity. By correlating the behavior of such twins—for example, by giving them intelligence tests or comparing their criminal or occupational records at different ages—it is possible to arrive at a number that expresses the degree of what statisticians call the variance in outcomes that is due to genes. The remainder is therefore due to environment. Behavior genetics also studies nonrelatives (that is, adopted siblings) raised in the same household. If the shared environment of family and upbringing are as powerful in molding behavior as the antihereditarians maintain, then such nonrelatives should exhibit a higher correlation of attributes than two randomly chosen nonrelatives. Comparing these two correlations then gives us a measure of the impact of shared environment.

The results of behavior genetics are frequently striking, showing strong correlations in the behavior of identical twins despite their having been raised by different parents with different cultural and/or socioeconomic backgrounds. This approach is not without its strong critics, however. The major problem has to do with what constitutes a different environment. In many cases, twins reared apart will nonetheless share many of the same environmental circumstances, making it impossible to disaggregate natural from cultural influences. Among the "shared environments" that a behavior geneticist may overlook is that of the mother's womb, which has a strong influence on how a given genotype develops into a phenotype, or individual human being. Identical twins necessarily share the same womb, but the same fetus growing up in a different womb might turn out quite differently if the mother is malnourished, drinks, or takes drugs.

The second, and less accurate, way of uncovering the natural sources of behavior is to do a cross-cultural survey of a particular trait or activity. By now we have a very large ethnographic record of behavior in a wide range of human societies, both those currently existing and those we know about through historical or archaeological records. If a characteristic appears in all or virtually all known societies, we can make a good, albeit circumstantial, case that it is due to genes rather than environment. This is the approach typically used in animal ethology, the comparative study of animal behavior.

One problem with this approach is that it is very difficult to find truly universal patterns in the way humans think and act. There is much more variability in human than in animal behavior, since human beings are to a much greater degree cultural creatures, learning how to behave from laws, customs, traditions, and other influences that are socially constructed rather than natural.[8] Post-Boas cultural anthropologists in particular have delighted in emphasizing the variability of human behavior. Many of the classics of twentieth-century anthropology have been those, like Margaret Mead's *Coming of Age in Samoa*, that purported to show that some cultural practice familiar in the West, like sexual jealousy or the regulation of adolescent female sexuality, was not practiced in some exotic non-Western culture.[9] This tradition lives on in countless "cultural studies" departments of universities around the United States, which emphasize deviant, transgressive, or otherwise unusual forms of behavior.

It nonetheless remains the fact that there are cultural universals: while particular forms of kinship, such as the Chinese five-generation family and the American nuclear family, are not universal, pair-bonding between males and females *is* a species-typical behavior for humans, in a way it is not for chimpanzees. The content of human languages is arbitrary and culturally constructed, but the "deep structures" of grammar first identified by Noam Chomsky, on which all languages are based, are not. Many of the examples of bizarre or atypical behavior used to undermine the existence of universal modes of cognition are, like Margaret Mead's study of Samoan adolescents, flawed. The Hopi Indians were said not to have a concept of time, when in fact they do; the anthropologist studying them simply didn't recognize it.[10] One might think colors would be good candidates for social construction, since what we identify as "blue" and "red" are in fact just points along a continuous spectrum of light wavelengths. And yet, one anthropological study asked members of widely disparate cultures to place colors used by their societies within a color table; it turned out that people perceived the same primary and secondary colors across cultural boundaries. This implies that there is something "hardwired" about color perception that is based on human biology, even if we do not know the specific genes or neurological structures that produce it.

Behavior genetics and cross-cultural anthropology begin with mac-

robehavior and make inferences about human nature based on correlations. The first begins with people who are genetically identical and
looks for environment-induced differences; the latter takes people
who are culturally heterogeneous and looks for genetically induced
similarities. Neither approach can ever fully prove its case to the satisfaction of critics, since both are based on a statistical inference,
with what are often large margins of error, and do not purport to de-
scribe the actual causal connections between genes and behavior.

This is all about to change. Biology can in theory supply information about the molecular pathways linking genes and behavior. Genes
control the expression—that is, the turning on and off—of other
genes, and they contain the code for the proteins that control chemical reactions within the body and are the building blocks of the body's
cells. Much of what we currently know about genetic causation is
limited to relatively simple single-gene disorders like Huntington's
chorea, Tay-Sachs disease, and cystic fibrosis, all of which can be
traced to a single allele (that is, a section of DNA that can vary between individuals). Higher-level behaviors, such as intelligence and
aggressiveness, are likely to have far more complex genetic roots, being the product of multiple genes that interact both with each other
and with the environment. But it seems almost inevitable that we will
know much more about genetic causation even if we never fully understand how behavior is formed.

For example, a gene for superior memory was inserted into a
mouse by Princeton biologist Joe Tsien. A brain cell component
known as the NDMA receptor has long been suspected of being
linked to the ability to form memories and is in turn the product of a
series of genes labeled NR1, NR2A, and NR2B. By performing a so-
called knockout experiment, in which a mouse was bred lacking the
NR1 gene, Tsien determined that the gene was indeed linked to
memory. In a second experiment, he added an NR2B gene to another
mouse and found that it in fact produced an animal with superior
memory.[11]

Tsien has not found a "gene for" intelligence; he has not even
found a "gene for" memory, given that memory is shaped by the interaction of many different genes. Intelligence itself is probably not just
one single characteristic but rather a collection of abilities that are af-

fected by a whole range of cognitive functions within the brain, of which memory is only one. But a piece of the puzzle is now in place, and more will come. It is obviously not possible to perform knockout gene experiments on human beings, but given the similarities between human and animal genotypes, it will become possible to make much stronger inferences about genetic causation than is currently feasible.

It is possible, moreover, to study differences in the distributions of different alleles and correlate them with population differences. We know, for example, that different population groups around the world have different distributions of blood types; roughly 40 percent of Europeans have type O blood, while Native Americans have almost exclusively type O.[12] The alleles that are linked to sickle-cell anemia are more common among African-Americans than among whites. The population geneticist Luigi Luca Cavalli-Sforza has mapped out a speculative history of past migrations of early humans as they wandered out of Africa to different parts of the globe, based on distributions of mitochondrial DNA (that is, DNA that is contained within the mitochondria, outside the cell nucleus, which is inherited from the mother's side).[13] He has gone further, linking these populations to the development of languages, and has provided a history of early language evolution in the absence of written records.

This kind of scientific knowledge, even in the absence of a technology that makes use of it, has important political implications. We have already seen this happen in the case of three higher-level behaviors with genetic roots—intelligence, crime, and sexuality—and there is much more to come.[14]

### The Heritability of Intelligence

In 1994, Charles Murray and Richard Herrnstein sparked a firestorm with the publication of their book, *The Bell Curve*.[15] Crammed with statistics and based heavily on a large data set, the National Longitudinal Survey of Youth, the book made two extremely controversial assertions. The first was that intelligence was largely inherited. In the language of statistics, Murray and Herrnstein argued that 60 to 70 percent of the variance in intelligence was due to genes, the rest to environmental factors such as nutrition, education, family structure, and the like. And second, they argued that genes played a role in the fact that African-Americans score lower than whites by about one standard deviation* on intelligence tests. Murray and Herrnstein maintained that in a world in which social barriers to mobility were falling and the rewards to intelligence rising, society would be increasingly stratified along cognitive lines. Genes and not social background would be the key to success. The most intelligent would walk away with most of the earnings; indeed, due to "assortative mating" (the tendency of people to marry like people) the cognitive elite would tend to increase its relative advantage over time. Those of lower intelligence faced severely limited life chances, and the ability of compensatory social programs to improve them was limited.[16] These arguments echoed those made earlier by psychologist Arthur Jensen in an article in the *Harvard Educational Review* that appeared in 1969, in which he came to similar pessimistic conclusions.[17]

It is no wonder that *The Bell Curve* produced such controversy. Murray and Herrnstein were denounced as racists and bigots.[18] In the words of one review, "As offensive and alarming as it might be, *The Bell Curve* . . . is simply another chapter in the continuing political economy of racism."[19] A common line of attack was to denounce the book's authors for being pseudoscientists whose findings were so shoddy and biased that they were not even worthy of serious argument, and to try to associate them with various skinhead and neo-Nazi organizations.[20]

But the book was only the latest salvo in an ongoing war between those who argue that intelligence has a high degree of heritability and those who argue that intelligence is largely shaped by environment. Conservatives are often sympathetic to arguments about natural human differences because they want to justify existing social hierarchies and oppose government intervention to correct them. The Left, by contrast, cannot abide the idea that there should be natural limits to the quest for social justice, and particularly that there are natural differences between human groups. The stakes in an issue like intel-

---

*A standard deviation is a statistical measure of how far a given population varies around a mean; approximately two thirds of a group will fall within one standard deviation above or below its mean.

ligence are so high that they have immediately spilled over into methodological disputes, with the Right arguing that cognitive ability was straightforward and measurable, and the Left maintaining that it was fuzzy and subject to gross mismeasurement.[21]

It is an uncomfortable fact that the development of modern statistics, and hence contemporary social science as a whole, is intimately bound up with psychometry and the work of a number of brilliant methodologists who also happened to be racists and eugenicists. First among them was Charles Darwin's cousin Francis Galton, coiner of the term *eugenics*, who in his book *Hereditary Genius* argued that exceptional ability tended to run in families.[22] Galton was one of the first people in the late nineteenth century to devise what he hoped would be an objective test to measure intelligence. He collected data systematically, and experimented with new mathematical methods for analyzing them.

Galton's disciple, Karl Pearson, was Galton Professor of Eugenics at University College, London, and a firm believer in social Darwinism who once wrote, "History shows me one way, and one way only, in which a high state of civilization has been produced, namely, the struggle of race with race, and the survival of the physically and mentally fitter race."[23] He also happened to be a superb methodologist and one of the founders of modern statistics. Every first-year student of statistics learns how to calculate "Pearson's $r$," the basic coefficient of correlation, and learns the chi square test for statistical significance, another of Pearson's inventions. Pearson developed the coefficient of correlation in part because he wanted to find a more accurate way of relating measurable phenomena, such as intelligence tests, to underlying biological characteristics, such as intelligence itself. (The Web page of the statistics department of University College proudly displays his accomplishments as an applied mathematician but discreetly ignores his writings on race and heredity.)

A third important methodologist was Charles Spearman, who invented the fundamental technique of factor analysis and the Spearman rank correlation, both indispensable statistical tools. Spearman, a psychometrician, noticed that tests for mental abilities were highly correlated with one another: if a person was good on a verbal test, for example, he or she also tended to be good on a math test. He postulated that there must be a general factor of intelligence, which he labeled g (for general intelligence), that was the underlying cause of an individual's performance on varied tests. Factor analysis developed out of his effort to isolate g in a rigorous way, and it remains central to contemporary discussions of heritable intelligence.

The association of psychometry with politically unpalatable views on race and eugenics may be enough to discredit the entire field for some, but what it in fact shows is that there is no necessary correlation between politically incorrect findings and bad science. Attacking the methodological credentials of people whose views one doesn't like and dismissing their work as "pseudoscience" is a convenient shortcut around arguing over substance. It was employed very effectively by the Left for much of the second half of the twentieth century, the high-water mark of this period being the publication in 1981 of Stephen Jay Gould's book *The Mismeasure of Man*.[24] Gould, a paleontologist with strong left-wing sympathies, began by picking such easy targets as Samuel George Morton and Paul Broca, nineteenth-century scientists who believed that intelligence could be inferred from the measurement of head size and whose faulty data were used in support of racist and anti-immigrant policies at the turn of the twentieth century. He then went on to attack more credible twentieth-century proponents of genetic theories of intelligence, such as Spearman and Sir Cyril Burt, on whom Arthur Jensen heavily relied.

The latter case was particularly notable because Burt, one of the giants of modern psychology, was charged in 1976 with deliberately falsifying data from studies of monozygotic twins to establish an estimate that intelligence was more than 70 percent a matter of heredity. Oliver Gillie, a British journalist, claimed in *The Sunday Times* of that year that Burt had made up coauthors and data and that his findings were a hoax. This gave tremendous ammunition to other critics, such as psychologist Leon Kamin, who argued that "there exist no data which should lead a prudent man to accept the hypothesis that IQ test scores are in any degree heritable."[25] He, along with Richard Lewontin and Steven Rose, went on to make a broad-based attack against the entire field of behavior genetics, which they regarded as a pseudoscience.[26]

Unfortunately, the idea that g refers to something real in the brain and that it has a genetic basis was not so easy to kill on methodological grounds alone. Later researchers, going back over Burt's work, demonstrated that the charges of deliberate fabrication were themselves fabricated.[27] In any event, Burt's studies were not the only ones of monozygotic twins that showed a high degree of heritability; there have been a number of others, including the 1990 Minnesota twin study, whose results are very similar to Burt's.

A serious and complex debate continues unabated among psychologists over the existence and nature of Spearman's g, with highly credible scholars making arguments on both sides.[28] From the moment it was first articulated in 1904, Spearman's theory that intelligence was a single thing has been attacked by those who believe that intelligence is in fact a collection of related abilities, each of which can vary within the same individual. One of the earliest proponents of this view was the American psychologist L. L. Thurstone; one of the more recent has been Howard Gardner, whose doctrine of "multiple intelligences" is widely known in American educational circles.[29] Defenders of the g factor point out that the argument is to some extent definitional: many of the abilities that Gardner labels intelligences, as Murray and Herrnstein themselves point out, might quite reasonably be called talents while preserving the term *intelligence* for a certain, more limited set of cognitive functions. They base their case for g on factor analysis and the strong statistical case that can be made that g is one thing. Critics make the reasonable counterargument that proponents of g are making an inference about the existence of an ability that, while it must have some physiological referent in the brain, no one has actually observed.

*The Bell Curve* led to the publication of a series of volumes by other psychologists and specialists on intelligence that summarized what is currently known about the link between intelligence and heredity.[30] It is clear from this literature that while many strongly disagree with Murray and Herrnstein on many of their central assertions, the issue they have identified—the importance of intelligence in modern societies and the implications of its having hereditary roots—is not going to go away. There is little disagreement, for example, that there is a substantial degree of heritability of whatever it is that intelligence tests measure, whether g or other, multiple factors of intelligence. A special issue of *American Psychologist*, published in the wake of *The Bell Curve*, summarized the consensus of the discipline as being that half of one's intelligence appears to be related to heredity as a child, and an even higher percentage as one becomes an adult.[31] There is a technical argument among specialists, concerning "broad" versus "narrow" heritability, that leads some to argue that the genetic component of intelligence is no more than about 40 percent,[32] but few take seriously the assertion by Kamin that there is no credible evidence linking performance on intelligence tests to heredity.

This difference in estimates of heritability has potentially important implications for public policy, because lower numbers, in the range of 40 to 50 percent, suggest that contrary to Murray and Herrnstein there are indeed environmental factors, which government policy might affect, that could help raise IQs. One can see the glass as half full rather than half empty: a better diet, education, a safe environment, and economic resources can all contribute to raising the 50 percent of a child's IQ that is due to environment and are therefore reasonable goals of social policy.

This environmental component also softens the blow with regard to the tortured issue of intelligence and race. The same special issue of *American Psychologist* confirmed that blacks do indeed score significantly lower on standardized intelligence tests than whites. The question is why. There are many circumstantial reasons to suggest that the gap is due much more to environmental than to genetic factors. A powerful one has to do with the so-called Flynn effect, named after psychologist James Flynn, who first noticed that IQ scores have been rising over the past generation in virtually every developed country.[33] It is extremely unlikely that this change is due to genetic factors, because genetic change does not occur this rapidly; Flynn himself is skeptical that people are on the whole that much smarter than they were a generation ago. This suggests that these massive gains in IQ are the result of some environmental factor that we understand only poorly, ranging from better nutrition (which has led the same populations to grow much taller over the same period as well) to education and the greater availability of mental stimulation. This suggests that socially disadvantaged groups, such as African-Americans, who have suffered relative disadvantages in diet, education, and other aspects of their social environment, will see their IQ scores rise over time as well. IQs of blacks have risen, just as those of Jews and other immigrant groups have, and the black-white gap has already narrowed to some degree; in the future, it may well fade to insignificance.

The point of this discussion of intelligence and genetics is not to argue in favor of one particular theory of intelligence over another, or for one specific estimate of the heritability of intelligence. My own observation of those around me (and particularly my own children) suggests that intelligence is not the work of a single g factor but is rather a series of closely related abilities. Commonsense observation also tells me that these abilities are heavily influenced by heredity. I suspect that further research on a molecular level is not going to lead to startling new findings about racial differences in intelligence. The amount of evolutionary time that has passed since the races separated is too short, and the degree of genetic variance between the races, when looking at characteristics that can be measured (such as the distribution of blood types), is too narrow to suggest that there can be strong group differences in this regard.

The issue is a different one. Even if we do not posit any breakthroughs in genetic engineering that will allow us to manipulate intelligence, the sheer accumulation of knowledge about genes and behavior will have political consequences. Some of these consequences may be very good: molecular biology may exonerate genes from responsibility for important differences between individuals or groups, just as Boas's research on head sizes debunked early-twentieth-century "scientific racism." On the other hand, the life sciences may give us news we would rather not hear. The political firestorm set off by *The Bell Curve* will not be the last on this subject, and the flames will be fed by further research in genetics, cognitive neuroscience, and molecular biology. Many on the Left would have liked simply to shout down arguments about genes and intelligence as inherently racist and the work of pseudoscientists, but the science itself will not permit this kind of shortcut. The accumulation of knowledge about molecular pathways to memory, such as that demonstrated by Joe Tsien's knockout-gene experiments on mice, will make future estimates of intelligence heritability far more precise. Brain imaging techniques, such as positron-emission tomography, functional resonance imaging, and magnetic resonance spectroscopy, are able to dynamically chart blood flow and neuron firings; correlating these with different kinds of mental activities may one day lay to rest with some finality the question of whether g is one thing or many things by localizing it in different parts of the brain. The fact that bad science in the past has been used for bad ends does not inoculate us against the possibility that good science in the future will serve only ends we deem good.

### Genetics and Crime

If there is anything more politically controversial than the link between heredity and intelligence, it is the genetic origins of crime. The effort to trace criminal behavior to biology has as long and problematic a history as psychometry, with research in this area suffering its share of bad methodology and links to the eugenics movement. The most famous discredited scientist in this tradition was the Italian physician Cesare Lombroso, who at the turn of the twentieth century examined both living and dead prisoners and developed a theory that there was a criminal physical type characterized by a sloping forehead, small head, and other characteristics. Under Darwin's influence, Lombroso believed that criminal "types" were throwbacks to an earlier stage of human evolution who somehow survived into the present. While Lombroso was responsible for the modern liberal view that certain people for biological reasons could not be held responsible for the crimes they committed, his work was so methodologically flawed that it thereafter came to be associated with phrenology and phlogistons in the annals of pseudoscience.[34]

Modern theories of the biological origins of crime come from the same source as modern theories of heredity and intelligence: behavior genetics. Any number of studies of monozygotic twins raised apart or nonrelatives raised together have produced correlations between genes and criminal behavior.[35] One particularly large study, based on a sample of 3,586 twins from the Danish Twin Register, showed that monozygotic twins had a 50 percent chance of sharing criminal behavior versus 21 percent for dizygotic (nonidentical) twins.[36] A large adoption study, again based on Danish data, compared monozygotic twins raised in households with criminal and noncriminal parents, against nonrelated siblings raised with and without criminal parents. The study showed that the criminality of a biological parent was a stronger predictor of criminal behavior in the child than the criminality of an adoptive parent, suggesting some form of genetic transmission of criminal propensities.

Academic critics of genetic theories of crime have made many of the same criticisms as with intelligence.[37] Twin studies often fail to detect subtle aspects of shared environment, fail to control for nongenetic factors that might be influencing correlations, or rely on surveys with small sample sizes. Travis Hirschi and Michael Gottfredson have argued that because crime is a socially constructed category, it cannot have biological origins.[38] That is, what one society counts as a crime is not necessarily illegal in another; how then can one speak of someone having a "gene for" date rape or loitering?

While many genetic theories of crime have been thoroughly discredited, crime is one area of social behavior where there are actually good reasons to think that genetic factors operate. Crime is of course a socially constructed category, but certain serious acts like murder and theft are not condoned in any society, and behavior traits, such as poor impulse control, that can lead certain individuals to transgress these rules could plausibly have genetic sources.[39] A criminal who shoots someone else in the head over a pair of running shoes is obviously not making a rational trade-off between short-term gratification and long-term costs; this can easily be the result of poor early childhood socialization, but it is not absurd to think that some people are simply innately bad at making this sort of decision.

If one moves from individual to group differences, it is possible to make a strong prima facie case for genetic factors in crime simply by observing that in virtually every known society and in every historical period, crimes have been overwhelmingly committed by young males, usually between the ages of 15 and 25.[40] Girls and women commit crimes, of course, as do elderly people, but there is something about adolescent males that particularly predisposes them to seek out physical self-assertion and to take risks in ways that make them transgress social rules. The biological anthropologist Richard Wrangham documented in his 1996 book *Demonic Males* the fact that male chimpanzees organize themselves into small groups that go out and deliberately ambush other male-bonded groups on the periphery of their colony's territory.[41] Given that human beings descended from a chimplike ancestor some 5 million years ago, and that there seems to be considerable continuity in human male proclivities for violence and aggression over this evolutionary period, the case for genetic causation would appear to be strong.[42]

A number of studies have suggested direct molecular pathways between genes and aggression. A late 1980s study of a Dutch family with a history of violent disorders traced the cause to genes that control the production of enzymes known as monoamine oxidases, or MAOs.[43] A later French study of mice showed that a similar defect in their MAO genes led them to turn extremely violent.[44]

Individuals can of course learn to control their impulses,[45] particularly if they are taught the proper habits at the right developmental stage.* Societies in turn can do a great deal to reinforce that self-control, and can deter and punish crime if self-control fails. These social factors account for the dramatically varying crime rates among societies (New York City at one point experienced more homicides in a year than the whole country of Japan) and within the same society over time.[46] But social control takes place in the context of biological impulses. The evolutionary psychologists Martin Daly and Margo Wilson have shown that homicide rates vary according to certain predictions of evolutionary biology—for example, that domestic homicide takes place much more frequently between nonkin (for instance, between husbands and wives or stepfathers and stepchildren) than between blood relatives.[47]

Whatever the exact trade-off between genes and environment with respect to crime, it is clear that any reasonable public discussion of this issue is politically impossible in the contemporary United States. The reason for this is that since African-Americans are disproportionately represented in the U.S. criminal population, any suggestion that there is a genetic component to crime is thought to imply that blacks are somehow genetically predisposed to be criminals. No serious academic researcher working on this issue has ever suggested anything of

---

*That impulse control, like language, is something that can be better learned at certain ages than at others is a further indication of the biological nature of crime.

the sort since the bad old days of scientific racism, but that has not prevented people from harboring deep suspicions that anyone even interested in this topic must have racist motives.

Such suspicions were fed in the early 1990s by Frederick K. Goodwin, a noted psychiatrist and head of the federal Alcohol, Drug Abuse, and Mental Health Administration. Goodwin, whom Tom Wolfe has described as "a certified yokel in the field of public relations," was describing the National Institute of Mental Health's Violence Initiative when he suggested that crime-ridden urban America was a "jungle."[48] Goodwin was evidently referring to a number of perfectly respectable studies that suggested that male violence is hardwired. Nonetheless, his inept choice of words led to immediate denunciations of him as a racist by Senator Edward Kennedy and Representative John Dingell, and condemnation of the Violence Initiative as a eugenics program designed to eliminate undesirables.

This set the stage for public protests organized around a conference titled "The Meaning and Significance of Research on Genetics and Criminal Behavior," organized by David Wasserman, a researcher at the University of Maryland, and funded in part by the National Institutes of Health's National Center for Human Genome Research.[49] The conference was scheduled, criticized, rescheduled, and finally held at a secluded location on the Chesapeake Bay in 1993. In response to pressure before the event, Wasserman sought to bring in critics of the field of genetics and crime and scheduled an entire panel on the history of the eugenics movement.[50] This did not prevent a number of conference participants from issuing a formal protest statement cautioning that "scientists as well as historians and sociologists must not allow themselves to be used to provide academic respectability for racist pseudoscience." The conference was disrupted by outside demonstrators chanting, "Maryland conference, you can't hide—we know you're pushing genocide!"[51] The likelihood that the National Institutes of Health (NIH) or National Institute of Mental Health will sponsor a similar event in the near future is, as one might imagine, low.

### Genes and Sexuality, Hetero and Homo

A third area in which accumulating knowledge about genetics has and will have important political implications is sexuality.[52] Few people would deny that sexuality has strong biological roots, and the case that many male-female differences are influenced by biology rather than by social environment is much stronger than that of racial differences. Human racial and ethnic groups (the borders between which are often fuzzy) have developed, after all, only over the past few tens of thousands of years—a mere blip in evolutionary time—while sexual differentiation has been around for hundreds of millions of years, long before there were human beings at all. Men and women differ physiologically, genetically (women, of course, having two X chromosomes, and men having an XY pair), and neurologically. It is a given of a certain important strand of contemporary feminism that all such sex differences end with the body, and that male and female minds are essentially identical. For people with this perspective, all sex differences become gender differences—that is, differences in the way boys and girls are socialized. But it is very implausible that this is wholly true, and an important branch of evolutionary biology has been arguing for the past generation that male and female minds have been shaped by differing requirements of evolutionary adaptation.[53]

There has been a great deal of empirical work on this subject over the past forty years. In 1974, psychologists Eleanor Maccoby and Carol Jacklin summarized much of what was then known in a massive volume entitled *Psychology of Sex Differences*.[54] That work debunked certain myths about how men and women differ—there is no credible evidence, for example, that boys and girls differ with regard to their sociableness, suggestibility, or analytic ability, or intelligence more generally. On the other hand, there were a number of areas in which a range of studies turned up consistent differences. Girls tend to have greater verbal ability than boys, boys excel in visual-spatial ability, boys have superior mathematical ability, and, finally, boys are far more aggressive.[55]

Maccoby's later book, *The Two Sexes*, shows that gender differentiation begins at a very early age. A wide variety of empirical studies show that boys' play is much more physical than girls', that they tend to establish better-defined dominance hierarchies than girls, that they are more competitive, and that their competitiveness tends to take place between groups rather than individuals. Boys are more physically aggressive than girls, though girls show greater relational aggression (that is, aggression through social ostracism or alienation). Boys' discourse is different, centering more often on aggressive-violent themes, while that of girls focuses on family relationships. And with respect to the choice of the sex of play partners in early childhood, boys and girls seem to be rigidly programmed to segregate themselves by sex.[56] Most of these results hold up cross-culturally. All of this suggests to Maccoby that there must be some biological element at work defining male and female behavior in addition to the socialization patterns to which they are conventionally attributed.[57]

When we get to the issue of genes and homosexuality, the political tables are almost completely turned. On the questions of genes and intelligence, genes and crime, and genes and sex difference, the Left is vehemently opposed to biological explanations and seeks to downplay any evidence that heredity exerts an important influence on any of these behaviors. On the question of homosexuality, the Left has made the opposite case: sexual orientation is not a matter of individual choice or social conditioning, but rather something given an individual as an accident of birth.

Homosexuality has always posed a particular problem for evolutionary biology. Since evolution is supposedly all about reproductive success, and homosexuals tend not to leave descendants, one would think that a gene for homosexuality would be eliminated from a population rather rapidly by natural selection. Contemporary evolutionary biologists have theorized that if a genetic factor produces homosexuality, it is the by-product of another, highly adaptive characteristic, one that possibly benefits females and is inherited on the mother's side.[58] It is believed that the brains of various animals, including humans, are sexualized at a prenatal stage by exposure to certain levels of various sex hormones that are genetically determined. Based on studies of mice, it has been hypothesized that male homosexuality is brought on by deficient exposure to prenatal testosterone.

Up to now the heritability of homosexuality has been estimated the same way as the heritability of intelligence or criminality, through twin and adoption studies. These studies have indicated rates of heritability from 31 to 74 percent in the case of men, and 27 to 76 percent in the case of women. A number of recent neuroanatomical studies have indicated that there are actually differences in the structure of three parts of the brain between homosexual and heterosexual men; according to Simon LeVay, differences show up particularly in the hypothalamus.[59] An actual genetic link between a certain spot on the X chromosome and homosexuality was in fact asserted by Dean Hamer, a researcher at the NIH.[60] Using standard genetic techniques of pedigree analysis of a group of self-acknowledged homosexual men, Hamer and his associates found a statistically significant correlation between sexual orientation and certain genetic markers on chromosomal region Xq28.

A number of critics have raised the same types of objections to this research as in the case of intelligence and crime.[61] Whatever the ultimate verdict on these theories, homosexuality, like male sexual selectivity, exists in virtually all known societies and would seem plausibly to have some natural basis. What is interesting is the *politics* of the issue. In contrast to intelligence and crime, where the Left attacked the very idea of heritability, many gay activists seized on the idea of the "gay gene" because the notion of genetic causation frees gays from moral responsibility for their condition. In this case, it had been the Right that argued that homosexuality was a lifestyle choice. The existence of a gay gene would "prove" that gayness was like freckles: a condition that no one could do anything about.

This argument makes no more sense than assertions that intelligence or criminality cannot be affected by environment. Apart from a few single-gene disorders like Huntington's chorea, genes are never 100 percent determinative of an individual's eventual condition,[62] and there is no reason to think that the existence of a gay gene means that culture, norms, opportunity, and other factors do not play a role in sexual orientation. The simple fact that there are many bisexuals indicates that there is a lot of plasticity in sexual orientation. If parents are worried that a camping trip with a gay scoutmaster might lead their son to have a homosexual experience, their son's lack of a gay gene is not going to relieve them of that anxiety.

On the other hand, people on the Right who feel that homosexuality is simply a matter of individual moral choice have to confront the same fact that those on the Left do with respect to intelligence or gender identity: nature imposes limits. Left-handed people can be taught to write or eat with their right hand, but it is always a struggle and never feels "natural" to them. In fact, homosexuality is no different from intelligence, criminality, or sexual identity insofar as it is a human predisposition that is partly determined by heredity, and partly conditioned by social environment and individual choice. One can argue in each case over the relative weights of the genetic and social causes, but the mere existence of a genetic factor makes discussion of these traits highly controversial because it suggests a limitation of moral agency and human potentiality.

One of the fondest hopes of twentieth-century social science was that the progress of the natural sciences would eliminate biology as a significant factor in human behavior. In many respects this hope proved true: there was no empirical basis for "scientific racism" because differences between racial or ethnic groups, or between men and women, proved to be much smaller than was believed in the immediate aftermath of Charles Darwin's theory of evolution. Mankind does in fact appear to be a remarkably homogeneous species, which supports our post-Enlightenment moral intuitions concerning the universal dignity of all people. But certain group differences—particularly between the sexes—remain. And biology still plays a major role in explaining differences between individuals within populations. The future accumulation of knowledge about human genetics is only going to increase our knowledge of these genetic sources of behavior, and therefore will continue to cause endless political controversy.

Scientific knowledge about causation will inevitably lead to a technological search for ways to manipulate that causality. For example, the existence of biological correlates of homosexuality—whether prenatal androgens, a distinctive neuroanatomy, or a gay gene on which the former are based—raises the possibility that there will one day be a "therapy" for gayness. And here the Left becomes quite justifiably queasy about its embrace of biological explanations, because these begin once again to threaten the equality of human dignity.

We can illustrate the problem by performing the following thought experiment. Assume that in twenty years we come to understand the genetics of homosexuality well and devise a way for parents to sharply reduce the likelihood that they will give birth to a gay child. This does not have to presuppose the existence of genetic engineering; it could simply be a pill that provided sufficient levels of testosterone in utero to masculinize the brain of the developing fetus. Suppose the treatment is cheap, effective, produces no significant side effects, and can be prescribed in the privacy of the obstetrician's office. Assume further that social norms have become totally accepting of homosexuality. How many expecting mothers would opt to take this pill?

My suspicion is that very many would, including people who today would be quite indignant at what they perceive to be antigay discrimination. They may perceive gayness to be something akin to baldness or shortness—not morally blameworthy, but nonetheless a less-than-optimal condition that, all other things being equal, one would rather have one's children avoid. (The desire of most people for descendants is one guarantee of this.) How then might this affect the status of gays, particularly those in the generation from which gayness was eliminated? Wouldn't this form of private eugenics make them more distinctive, and greater targets for discrimination, than they were before? More important, is it obvious that the human race would be improved if gayness were eliminated from it? And if it is not obvious, should we be indifferent to the fact that these eugenic choices are being made, so long as they are made by parents rather than coercive states?

**3**

# NEUROPHARMACOLOGY AND THE CONTROL OF BEHAVIOR

"Becoming sick and harboring suspicion are sinful to them: one proceeds carefully. A fool, whoever still stumbles over stones or human beings! A little poison for now and then: that makes for agreeable dreams. And much poison in the end, for an agreeable death."

Friedrich Nietzsche, *Thus Spoke Zarathustra*, I.5

The thinker whose work suffered perhaps the greatest rise and fall from grace in the twentieth century was the founder of psychoanalysis, Sigmund Freud. At midcentury, Freud was universally esteemed in the West as the man who had uncovered the deepest truths about human motivation and desire. The Oedipus complex, the unconscious, penis envy, the death wish—Freud's concepts were thrown around at cocktail parties by cognoscenti who wanted to prove their sophistication. But by the end of the century, Freud was regarded by most of the medical profession as little more than an interesting footnote in intellectual history, someone who was more a philosophizer than a scientist. For this we have to thank advances in cognitive neuroscience and the new field of neuropharmacology.

Freudianism was built on the premise that mental illness, including serious diseases such as manic depression and schizophrenia, was primarily psychological in nature—the result of mental dysfunctions that occurred somewhere above the biological substrate of the brain. This view was undermined by the drug lithium, which was serendipitously discovered by an Australian psychiatrist, John Cade, when he administered it to manic-depressive mental patients in 1949.[1] A number of these patients were miraculously cured, beginning a process that would see Freudian "talk" therapy replaced almost entirely over the next two generations by drug therapy. Lithium was only the beginning of an explosive period of research and development in neuropharmacology, which would lead by the end of the century to a new generation of drugs, like Prozac and Ritalin, whose social impact we are only now beginning to understand.

The rise of psychotropic drugs has coincided with what has been called the neurotransmitter revolution—that is, a vast increase in scientific knowledge about the biochemical nature of the brain and its mental processes.[2] Freudianism might be compared to the theory developed by a group of primitive tribesmen who found a working automobile and tried to explain its internal functioning without being able to open the hood. They would observe the strong correlation between stepping on the gas pedal and moving forward, and would theorize that there was some device connecting the two that converted a liquid into the motion of the wheels—perhaps a large squirrel in a cage or a homunculus of some sort. But they would understand nothing about hydrocarbons, internal combustion, or the valves and pistons that did the actual energy conversion.

Modern neuroscience has, in effect, lifted the hood and permitted us to peer, however tentatively, at the engine. The dozen or so neurotransmitters, such as serotonin, dopamine, and norepinephrine, control the firing of nerve synapses and the transmission of signals across the neurons in the brain. The levels of these neurotransmitters and the way they interact directly affect our subjective feelings of well-being, self-esteem, fear, and the like. Their baseline levels are affected by things that go on in the environment and are very much related to what we understand to be personality. Long before genetic engineering becomes a possibility, knowledge of brain chemistry and the ability to manipulate it will become an important source of behavior control that will have significant political implications. We are already in the midst of this revolution and do not have to spin out science fiction scenarios to see how it might unfold.

Take the antidepressant Prozac, manufactured by Eli Lilly, and related drugs, such as Pfizer's Zoloft and SmithKline Beecham's Paxil. Prozac, or fluoxetine, is a so-called selective serotonin reuptake inhibitor (SSRI), which, as its name implies, blocks the reabsorption of serotonin by the nerve synapses and effectively increases the levels of serotonin in the brain. Serotonin is a key neurotransmitter: low levels are associated, in both humans and other primates, with poor impulse control and uncontrolled aggression against inappropriate targets, and in humans, with depression, aggression, and suicide.[3]

It is unsurprising, then, that Prozac and its relatives have emerged as a major cultural phenomenon in the late twentieth century. Peter D. Kramer's *Listening to Prozac* and Elizabeth Wurtzel's *Prozac Nation* both celebrate Prozac as a wonder drug that effects miraculous changes in personality.[4] Kramer describes a patient of his, Tess, who was chronically depressed, locked into a series of masochistic relationships with married men, and at a dead end at work. Within weeks of taking Prozac, her personality changed completely: she dropped her abusive relationship and started dating other men, changed her circle of friends entirely, and became more confident and less conciliatory in her management style at work.[5] Kramer's book became a best-seller and contributed enormously to the use and acceptance of the drug. Today, Prozac and its relatives have been taken by some 28 million Americans, or 10 percent of the entire population.[6] Because more women than men suffer from depression and low self-esteem, it has also become something of a feminist icon: Tess's success in breaking free of a demeaning relationship has been repeated, evidently, by many of the women for whom serotonin reuptake inhibitors have been prescribed.

It is not surprising that drugs reputed to have these kinds of effects have generated substantial controversy. Some studies have indicated that Prozac is not as effective as claimed,[7] and Kramer has been criticized for greatly exaggerating its impact. By far the larger portion of anti-Prozac literature has consisted of books like Peter Breggin and

Ginger Ross Breggin's *Talking Back to Prozac*[8] and Joseph Glenmullen's *Prozac Backlash,*[9] which argue that Prozac has a host of side
effects that its manufacturer has tried to cover up. These critics have
argued that Prozac is responsible for weight gain, disfiguring tics,
memory loss, sexual dysfunction, suicide, violence, and brain damage.

It may well be that in time, Prozac will go the way of the antipsy-
chotic Thorazine and will no longer be regarded as a wonder drug because of long-term side effects that were poorly understood when it
was first introduced. But the more difficult political and moral problem will occur if Prozac is found to be completely safe and if it, or
similar drugs yet to be discovered, work just as advertised. For Prozac
is said to affect that most central of political emotions, the feeling of
self-worth, or self-esteem.

Self-esteem is of course a trendy psychological concept, something Americans are constantly being told they need more of. But it
refers to a critical aspect of human psychology, the desire all people
have for recognition. Socrates, in Plato's *Republic*, argued that there
are three distinct parts of the soul, a desiring part, a rational part, and
what he labeled *thymos,* a Greek word usually translated as "spirited-
ness." *Thymos* is the prideful side of the human personality, the part
that demands that other people recognize one's worth or dignity. It is
not a desire for some material good or object to satisfy a need—the
"utility" that economists usually understand as the source of human
motivation—but rather an intersubjective demand that some other
human being acknowledge one's status. Indeed, the economist Robert
Frank points out that much of what we understand to be economic
interest is really a demand for status recognition, or what he labels po-
sitional goods.[10] That is, we want that Jaguar not so much because we
love beautiful cars but because we want to trump our neighbor's
BMW. The demand for recognition does not have to be personal; one
can demand that other people recognize one's gods, or sense of the sacred, or nation, or just cause as well.[11]

Most political theorists have recognized the centrality of recognition and the way that it is particularly crucial to politics. A prince
fighting another prince doesn't need the land or money; he usually has
more than he knows what to do with. What he wants is recognition of
his dominion or sovereignty, the acknowledgment that he is king of

kings. The demand for recognition frequently trumps economic interest: new nations like Ukraine and Slovakia might have been better off
remaining parts of larger countries, but what they sought was not economic welfare but their own flag and seat at the United Nations. It is
for this reason that the philosopher Hegel believed that the historical
process was fundamentally driven by the struggle for recognition, beginning with a primordial "bloody battle" between two contestants for
who would be master and who would be slave, and ending in the
emergence of modern democracy, in which all citizens were recognized as being free and worthy of equal recognition.

Hegel believed that the struggle for recognition was a purely human phenomenon—indeed, that it was in some sense central to what
it meant to be a human being. But in this he was wrong: there is a biological substrate for the human desire for recognition that is present
in a number of other animal species. Members of many species sort
themselves out into dominance hierarchies (the term *pecking order*
comes, of course, from chickens). When one gets to mankind's primate relatives, such as gorillas and, particularly, chimpanzees, the
struggle for status within a dominance hierarchy begins to look very
human indeed. The primatologist Frans de Waal has described at
great length the struggles for status that took place within a captive
chimp colony in the Netherlands in his appropriately named book
*Chimpanzee Politics.*[12] Male chimps form coalitions, scheme and betray one another, and evidently feel emotions that look very much like
pride and anger when their rank within the colony is or is not recognized by their fellows.

The human struggle for recognition is, of course, infinitely more
complex than what takes place among animals. Human beings, with
their memory, learning, and enormous capacity for abstract reasoning,
are able to direct the struggle for recognition to ideologies, religious
beliefs, tenure at universities, Nobel Prizes, and myriad other honors.
What is significant, however, is that the desire for recognition has a
biological basis and that that basis is related to levels of serotonin in
the brain. It has been shown that monkeys at the low end of the dominance hierarchy have low levels of serotonin and that, conversely,
when a monkey wins alpha male status, he feels a "serotonin high."[13]

It is for this reason that a drug like Prozac looks so politically consequential. Hegel argues, with some justice, that the entire human historical process has been driven by a series of repeated struggles for recognition. Virtually all human progress has been the by-product of the fact that people were never satisfied with the recognition they received; it was through struggle and work alone that people could achieve it. Status, in other words, had to be earned, whether by kings and princes, or by your cousin Mel, seeking to rise to the rank of shop foreman. The normal, and morally acceptable, way of overcoming low self-esteem was to struggle with oneself and with others, to work hard, to endure sometimes painful sacrifices, and finally to rise and be seen as having done so. The problem with self-esteem as it is understood in American pop psychology is that it becomes an entitlement, something everyone needs to have whether it is deserved or not. This devalues self-esteem and makes the quest for it self-defeating.

But now along comes the American pharmaceutical industry, which through drugs like Zoloft and Prozac can provide self-esteem in a bottle by elevating brain serotonin. The ability to manipulate personality in the way Peter Kramer describes raises some interesting questions. Could all of that struggle in human history have been avoided if only people had had more serotonin in their brains? Would Caesar or Napoleon have felt the need to conquer most of Europe if he had been able to pop a Prozac tablet every now and then? If so, what would have become of history?

There are clearly millions of people in the world who are clinically depressed and whose feelings of self-worth fall far below what they should be. For them, Prozac and related drugs have been a godsend. But low levels of serotonin do not demarcate a clear pathological condition, and the existence of Prozac opens the way for what Kramer famously labeled cosmetic pharmacology: that is, the taking of a drug not for its therapeutic value but simply because it makes one feel "better than good." If a sense of self-esteem is so crucial to human happiness, who would not want more of it? And so the path is opened toward a drug that in certain ways looks uncomfortably like the soma of Aldous Huxley's *Brave New World*.

If Prozac appears to be some type of happiness pill, Ritalin has come to play the role of an overt instrument of social control. Ritalin[14] is the trade name of methylphenidate, a stimulant closely related to methamphetamine, the street drug known in the 1960s as speed. It is used today to treat a syndrome known as attention deficit–hyperactivity disorder, or ADHD, a "disease" commonly associated with young boys who have trouble sitting still in class.

Attention deficit disorder (ADD) was first listed as a disease in 1980 in the American Psychiatric Association's *Diagnostic and Statistical Manual of Mental Disorders* (*DSM*), the bible of official mental diseases. The name of the disease was changed in a later edition of the *DSM* to attention deficit–hyperactivity disorder, *hyperactivity* being added as a qualifying characteristic. The entry of ADD and then ADHD into the *DSM* was in itself an interesting development. Despite several decades of searching, no one has been able to identify a cause of ADD/ADHD. It is a pathology recognized only by its symptoms. The *DSM* lists a number of diagnostic criteria for the disease, such as trouble concentrating and overactivity in motor functions. Doctors make what amounts to an often highly subjective diagnosis if the patient exhibits enough of the listed symptoms, whose very existence may often not be open-and-shut.[15]

It is thus not surprising that psychiatrists Edward Hallowell and John Ratey assert in their book *Driven to Distraction,* "Once you catch on to what this syndrome is about, you'll see it everywhere."[16] By their account, 15 million Americans may be suffering from some form of ADHD. If this is true, then the United States is experiencing an epidemic of truly staggering proportions.

There is of course a simpler explanation, which is that ADHD isn't a disease at all but rather just the tail of the bell curve describing the distribution of perfectly normal behavior.[17] Young human beings, and particularly young boys, were not designed by evolution to sit around at a desk for hours at a time paying attention to a teacher, but rather to run and play and do other physically active things. The fact that we increasingly demand that they sit still in classrooms, or that parents and teachers have less time to spend with them on interesting tasks, is what creates the impression that there is a growing disease. In the words of Lawrence Diller, a doctor and the author of a critique of Ritalin:

> We are left with the possibility that ADD may be a catch-all condition encompassing a variety of children's behavioral problems with various causes, both biologically predetermined and psychosocial.

And the fact that Ritalin helps with so many problems may be encouraging the ADD diagnosis to expand its boundaries.[18]

Ritalin is a central nervous system stimulant that is chemically related to such controlled substances as methamphetamine and cocaine. Its pharmacological effects are very similar to those of the latter drugs, increasing attention span, creating a sense of euphoria, building short-term energy levels, and allowing greater focus. Indeed, lab animals given the option of self-administering either Ritalin or cocaine do not show a strong preference for one over the other. These drugs will increase the focus, concentration, and energy levels of normal people as well. If used to excess, Ritalin can have side effects similar to those of methamphetamine and cocaine, including insomnia and weight loss. This is why doctors prescribing Ritalin to children recommend periodic "drug holidays." In the low dosages usually prescribed for children, Ritalin does not appear to be anywhere as intensely addictive as cocaine, but in higher dosages its effects can be similar. This had led the U.S. Drug Enforcement Administration (DEA) to classify it as a Schedule II drug, requiring prescription in triplicate by physicians, and controls on how much of the drug can be produced overall.[19]

The beneficial psychological effects of Ritalin explain why it is used—or, as the DEA would have it, abused—by increasing numbers of people who are not diagnosed with ADHD. According to Diller, "The drug potentially improves the performance of anyone—child or adult, ADD-diagnosed or not."[20] During the 1990s, Ritalin became one of the fastest-growing drugs used in high schools and on college campuses, as students discovered it helped them study for exams and pay better attention in class. According to a doctor at the University of Wisconsin, "The study rooms are as good as some of the local pharmacies here."[21] Elizabeth Wurtzel of Prozac fame describes chopping up and snorting forty Ritalin pills a day, which led to emergency room visits and detoxification therapy, at which she met mothers who stole their children's pills for their own use.[22]

The politics of Ritalin is very revealing of the impoverished thought categories by which we have come to understand character and behavior, and it offers us a foretaste of what will come if and when genetic engineering, with its potentially far more powerful behavioral enhancements, becomes available. Those who believe that they are suffering from ADHD are often desperate to believe that their inability to concentrate or to perform well in some life function is not, as they have often been told, a matter of poor character or weak will but the result of a neurological condition. Like gays who point to a "gay gene" as the source of their behavior, they would like to absolve themselves of personal responsibility for their actions. As the title of one popular recent pro-Ritalin book puts it, "it's nobody's fault."[23]

Now, it is certainly the case that there are many people whose hyperactivity or inability to concentrate is so extreme that one would grant that biology is the primary determinant of their behavior. But what about people who find themselves in, say, the fifteenth percentile of the normal distribution for attentiveness? There is some biological basis for their condition, but clearly they can do things that would affect their final degree of attentiveness or hyperactivity. Training, character, determination, and environment more generally would all play important roles. To classify people in this situation as suffering from a pathology is therefore to blur the line between therapy and enhancement. Yet this is exactly the demand that proponents of the medicalization of ADHD have made.

In this they are supported by some very important interests.[24] First and foremost is the simple self-interest of parents and teachers who do not want to spend the time and energy necessary to discipline, divert, entertain, or train difficult children the old-fashioned way. It is understandable, of course, that harried parents and overworked teachers should want to make life easier for themselves by taking a medical shortcut, but what is understandable does not always amount to the same thing as what is right. The most important lobby representing these interests in the United States is CHADD, or Children and Adults with Attention-Deficit/Hyperactivity Disorder, a nonprofit self-help group founded in 1987 and composed mostly of parents with children diagnosed as having ADHD. CHADD sees itself as a support group and clearinghouse for the most up-to-date information on ADHD and its treatment, and it has lobbied hard to have ADHD classified as a disability and to see that ADHD-diagnosed children qualify for special education under the Individuals with Disabilities Education Act (IDEA).[25] CHADD has been particularly concerned that victims of ADHD not be stigmatized for their condition. In 1995 it launched a huge campaign to have Ritalin reclassified as a Schedule III drug, which would lift DEA controls on overall production and considerably relax the conditions under which it is prescribed and obtained.[26]

The second important source of support for the medicalization of ADHD is the pharmaceutical industry, and particularly companies, like Novartis (formerly Ciba-Geigy), that manufacture Ritalin and its relatives. Eli Lilly, the manufacturer of Prozac, has spent a fortune beating back negative stories about the side effects of its major revenue source, and the same has been true of Novartis. Novartis lobbied strongly to have Ritalin reclassified as a Schedule III drug and built pressure to have production quotas rapidly raised by spreading stories in the early 1990s of an impending production shortage. In 1995 the drug company overreached itself, however, when the reclassification effort collapsed in the wake of news that Novartis had failed to disclose donations to CHADD of nearly $900,000.

The medicalization of a condition like ADHD has important legal and political consequences. Under U.S. law, ADHD is currently classified as a disability and as such wins its victims coverage under two separate laws, Section 504 of the Vocational Rehabilitation Act of 1973, and the Individuals with Disabilities Education Act, which was passed in 1990. The former forbids discrimination against people with disabilities; the latter provides extra funding for special education of those with officially recognized educational disabilities. The addition of ADHD to IDEA's disabilities list was the result of a protracted political battle that pitted CHADD and other medical and advocacy groups against the National Education Association (NEA)—the national teachers' union—and the National Association for the Advancement of Colored People (NAACP). The NEA disliked the budgetary consequences of an expansive list of disabilities, while the NAACP worried that black children might be more easily classified as having learning disabilities and then medicated than white children. ADHD was finally put on the official disabilities list in 1991 after an intensive letter-writing and lobbying campaign on the part of CHADD and other parents' groups.[27]

As a consequence of ADHD's classification as an official disability, children with the condition are entitled to special education services in school districts around the United States. ADHD students can demand extra time to take standardized tests, a practice that schools have acceded to in order to avoid being sued. According to *Forbes* magazine, the Whittier Law School was sued by an ADD-diagnosed student for providing only twenty minutes of extra time per hour-long exam to complete it. Rather than risk litigation, the school settled.[28]

Many conservatives have complained about the expansiveness of current American definitions of disabilities under IDEA on the basis of cost. But the more important objection is a moral one: by classifying ADHD as a disability, society has in effect taken a condition with both biological *and* psychosocial causes and said that biology should predominate. Individuals who in fact have some degree of control over their behavior are told that they do not, and the nondisabled part of society then reallocates resources and time to see that they are compensated for something that is actually at least partly under their control.

The concern of groups such as the NAACP that psychotropic drugs like Ritalin might be used disproportionately in minority communities might have some validity as well. In the United States, there has been a remarkable increase in the number of prescriptions for psychotropic drugs (primarily but not exclusively Ritalin and its relatives) being given to extremely young children (that is, preschoolers or even younger) for behavioral problems. A 1998 study showed that among Michigan Medicaid recipients, some 57 percent of those under the age of 4 who were diagnosed with ADHD were being prescribed one or more psychotropic medications.[29] One particular study, which caused a small political furor when it was released, showed that in 1995 stimulants were being given to more than 12 percent of 2- to 4-year-olds in one large Midwestern Medicaid program, and antidepressants to nearly 4 percent. Reading between the lines of the study, it was clear that these drugs were being prescribed at significantly higher rates in the heavily minority Medicaid programs than they were in the better-off HMO also under study.[30]

There is a disconcerting symmetry between Prozac and Ritalin. The former is prescribed heavily for depressed women lacking in self-esteem; it gives them more of the alpha-male feeling that comes with high serotonin levels. Ritalin, on the other hand, is prescribed largely for young boys who do not want to sit still in class because nature never designed them to behave that way. Together, the two sexes are gently nudged toward that androgynous median personality, self-satisfied and socially compliant, that is the current politically correct outcome in American society.

The second, neuropharmacological wave of the biotech revolution has already come crashing down around us. It has already produced a pill that looks like soma and a pill for socially controlling children, pills that appear to be far more effective than early childhood social-ization and Freudian talk therapies of the twentieth century ever were. Their use has spread to millions and millions of people around the world, with much controversy over their potential long-term health consequences for the body, but almost no argument over what they imply about conventional understanding of identity and moral behavior.

Prozac and Ritalin are only the first generation of psychotropic drugs. In the future, virtually everything that the popular imagination envisions genetic engineering accomplishing is much more likely to be accomplished sooner through neuropharmacology.[31] A class of drugs known as benzodiazepines may be used to affect the gamma-aminobutyric acid (GABA) system, to reduce anxiety, help maintain restful but active wakefulness, and produce adequate sleep in a shorter period, without the side effects of sedation. Acetylcholine sys-tem enhancers may be used to improve the ability to learn new facts, retain knowledge, and improve factual recall. Dopamine system en-hancers may be used to increase stamina and motivation. Selective serotonin reuptake inhibitors in combination with drugs that affect the dopamine and norepinephrine systems may produce behavioral changes in areas in which the different neurotransmitter systems in-teract. Finally, it may be possible to manipulate the endogenous opi-ate system to decrease sensitivity to pain and increase the threshold of pleasure.

We do not have to wait for genetic engineering and designer ba-bies to have a foretaste of the kinds of political forces that will push forward new medical technologies; we can see all of them operating in the realm of neuropharmacology. The spread of psychotropic drugs in the United States demonstrates three powerful political trends that will reappear with genetic engineering. The first is the desire on the part of ordinary people to medicalize as much of their behavior as possible and thereby reduce their responsibility for their own actions. The second is the pressure of powerful economic interests to assist in this process. These interests include social service providers such as teachers and doctors, who will always prefer biological shortcuts to complex behavioral interventions, as well as the pharmaceutical companies that produce the drugs. The third trend, which flows from the attempt to medicalize everything, is the tendency to expand the therapeutic realm to cover an ever larger number of conditions. It will always be possible to get a doctor somewhere to agree that someone's unpleasant or distressing situation constitutes a pathology, and it is only a matter of time before the larger community comes to regard such a condition as a legal disability subject to compensatory public intervention.

I have spent this much time on drugs like Prozac and Ritalin not because I believe they are inherently evil or harmful, but rather because I believe they are harbingers of things to come. It may be that in a few years they will fall out of favor because of unanticipated side effects. But if so, they will simply be replaced in time by yet more sophisticated psychotropic drugs with more powerful and targeted effects.

The term *social control*, of course, conjures up right-wing fantasies of governments using mind-altering drugs to produce compliant subjects. That particular fear would seem to be misplaced for the foreseeable future. But social control is something that can be exercised by social players other than the state—by parents, teachers, school systems, and others with vested interests in how people behave. Democracies, as Alexis de Tocqueville pointed out, are subject to a "tyranny of the majority," in which popular opinions crowd out genuine diversity and difference. In our day this has come to be known as political correctness, and it is worthwhile worrying about whether modern biotechnology will not soon be in the business of providing powerful new biological shortcuts to the reaching of politically correct ends.

Neuropharmacology also points the way to possible political responses. There is no question that drugs like Prozac and Ritalin help enormous numbers of people who could not be helped in other ways. This is so because there are in fact many severely depressed or excessively hyperactive people whose biological condition prevents them from enjoying what most people would regard as a normal life. Apart from, perhaps, Scientologists, there are few people who would want an outright ban on such medicines or who would choose to curtail their use in cases that are clearly therapeutic. What can and should trouble us is the use of such drugs either for "cosmetic pharmacology," to enhance otherwise normal behavior, or to exchange one normal behavior in favor of another that someone thinks is socially preferable.

American society, like most others, embodies these reservations in its drug laws. But our laws are frequently inconsistent and poorly thought out, not to mention poorly enforced. Take the drug Ecstasy, the street name for MDMA, or methylenedioxymethamphetamine, one of the fastest-growing illicit drugs of the 1990s. Ecstasy, a stimulant very similar to methamphetamine, became the rage in dance clubs. According to the U.S. National Institute on Drug Abuse, 8 percent of all twelfth graders, or 3.4 million Americans, have used MDMA at least once in their lifetime.[32]

While chemically related to Ritalin, Ecstasy has an effect more like that of Prozac: it stimulates the release of serotonin in the brain. The effects of Ecstasy are powerfully mood- and personality-altering, just as in the case of Prozac. Consider the following story of one Ecstasy user:

> Users consistently describe that initial high as one of the greatest experiences of their lives. Jennie, 20, is a college student who lives in upstate New York. We met during a December visit she made to Washington. She has the delicate features and fair complexion of a folk-music princess. The first time she took Ecstasy, she told me, was a year ago. It inspired deep reflections. "I decided that one day I'd have children," she said, with striking frankness. "Before, I really did not think I was going to have children. I didn't think I'd be a very good mother, because I'd been kind of physically and mentally abused by my father. But then I realized, 'I'm going to love my children and I'm going to take care of them,' and my decision didn't

change afterward." She also says that on her first Ecstasy trip, she began to forgive her father, realizing that "there's no such thing as a bad person."[33]

Other descriptions of Ecstasy make it sound like a drug that heightens social sensitivity, promotes human bonding, and increases focus—all effects that generally receive approbation from society and are eerily similar to those attributed to Prozac. And yet Ecstasy is a controlled substance whose sale and use is illegal under all circumstances in the United States, whereas Ritalin and Prozac are drugs that may be legally prescribed by a physician. What accounts for the difference?

One obvious answer is that Ecstasy harms the body in ways that Ritalin and Prozac arguably do not. The National Institute on Drug Abuse's Web page on Ecstasy states that the drug induces psychological problems such as "confusion, depression, sleep problems, drug craving, severe anxiety, and paranoia"; physical symptoms such as "muscle tension, involuntary teeth clenching, nausea, blurred vision, rapid eye movement, faintness, and chills or sweating"; and has been shown to produce permanent brain damage in monkeys.

The literature on Ritalin and Prozac, in fact, is full of anecdotal evidence of similar kinds of side effects (with the exception of permanent brain damage in monkeys) from these legal drugs. Some have argued that the difference is largely a matter of dosage: if abused, Ritalin can also produce severe side effects, which is why it can be taken only under the guidance of a physician. But this begs the question: Why not legalize Ecstasy as a Schedule II drug? Or alternatively, why not search for a pharmacologically similar drug that minimizes Ecstasy's side effects?

The answer to this question gets at the heart of our confusion over the criminalization of drugs. We feel very ambivalent about substances that have no clear therapeutic purpose, and whose only effect is to make people feel good. We feel particularly ambivalent if the high produced by the drug seriously impairs the user's ability to function normally, as is the case with heroin and cocaine. But we also find it hard to justify our ambivalence, since doing so involves making judgments as to what a person's "normal functioning" is. How can we justify banning marijuana when alcohol and nicotine, two other drugs that make us feel good, are legal?* In light of these difficulties, we find it much easier to ban drugs on the basis of clear harms to the body—that they are addictive, that they cause physical impairment, that they lead to long-term unwanted side effects, and the like.

We are, in other words, unwilling to take a clear stand on drugs solely on the basis that they are bad for the soul—or, in contemporary medical language, on the basis of psychological effects alone. If tomorrow a pharmaceutical company invented an honest-to-God Huxleyan soma tablet that made you happy and socially bonded, without any harmful side effects, it is not clear that anyone could articulate a reason people shouldn't be allowed to take it. There are many libertarians on both the Right and the Left who argue that we should stop worrying about other people's souls or internal states altogether, and let people enjoy whatever drugs they choose as long as they don't harm anyone else. If a cranky traditionalist objected that this soma wasn't therapeutic, the psychiatric profession could probably be depended on to declare unhappiness a pathology and to put it into the *DSM* next to ADHD.

So we don't have to await the arrival of human genetic engineering to foresee a time when we will be able to enhance intelligence, memory, emotional sensitivity, and sexuality, as well as reduce aggressiveness and manipulate behavior in a host of other ways. The issue has already been joined with the current generation of psychotropic drugs, and will be put into much sharper relief with those shortly to come.

---

*I believe it is possible to distinguish between alcohol and nicotine, on the one hand, and a drug like marijuana, on the other, in terms of psychological effect. It is possible to drink and smoke moderately in ways that do not impair one's general social functioning; indeed, many people believe that moderate drinking is a boon to sociability. Other drugs, however, produce a high that is incompatible with any kind of normal social functioning.

**4**

## THE PROLONGATION OF LIFE

**Many die too late, and a few die too early. The doctrine sounds strange: "Die at the right time!"**

**Die at the right time—thus teaches Zarathustra. Of course, how could those who never live at the right time die at the right time? Would that they had never been born! Thus I counsel the superfluous. But even the superfluous still make a fuss about their dying; and even the hollowest nut wants to be cracked.**

**Friedrich Nietzsche, *Thus Spoke Zarathustra*, I.21**

The third pathway by which contemporary biotechnology will affect politics is through the prolongation of life, and the demographic and social changes that will occur as a result. One of the greatest achievements of twentieth-century medicine in the United States was the raising of life expectancies at birth from 48.3 years for men and 46.3 for women in 1900 to 74.2 for men and 79.9 for women in 2000.[1] This shift, coupled with dramatically falling birthrates in much of the developed world, has already produced a very different global demographic backdrop for world politics, whose effects are arguably being felt already. Based on birth and mortality patterns already in place, the world will look substantially different in the year 2050 than it does today, even if biomedicine fails to raise life expectancies by a single year over that period. The likelihood that there will not be significant advances in the prolongation of life in this period is small, however, and there is some possibility that biotechnology will lead to very dramatic changes.

One of the areas most affected by advances in molecular biology has been gerontology, the study of aging. There are at present a number of competing theories as to why people grow old and eventually die, with no firm consensus as to the ultimate reasons or mechanisms by which this occurs.[2] One stream of theory comes out of evolutionary biology and holds, broadly, that organisms age and die because there are few forces of natural selection that favor the survival of individuals past the age at which they are able to reproduce.[3] Certain genes may favor an individual's ability to reproduce but become dysfunctional at later periods of life. For evolutionary biologists, the big mystery is not why individuals die but why, for example, human females have a long postmenopausal life span. Whatever the explanation, they tend to believe that aging is the result of the interaction of a large number of genes, and that therefore there are no genetic shortcuts to the postponement of death.[4]

Another stream of theory on aging comes out of molecular biology and concerns the specific cellular mechanisms by which the body loses its functionality and dies. There are two types of human cells: germ cells, which are contained in the female ovum and male sperm, and somatic cells, which include the other hundred trillion or so cells that constitute the rest of the body. All cells replicate by cell division. In 1961, Leonard Hayflick discovered that somatic cells had an upper limit in the total number of divisions they could undergo. The number of possible cell divisions decreased with the age of the cell.

There are a number of theories as to why the so-called Hayflick limit exits. The leading one has to do with the accumulation of random genetic damage as cells replicate.[5] With each cellular division, environmental factors like smoke and radiation, as well as chemicals known as free hydroxyl radicals and cellular waste products, can prevent the accurate copying of the DNA from one cell generation to the next. The body has a number of DNA repair enzymes that oversee the copying process and fix transcription problems as they arise, but these fail to catch all mistakes. With continued cell replication, the DNA

damage builds up in the cells, leading to faulty protein sy
impaired functioning. These impairments are in turn the
eases characteristic of aging, such as arteriosclerosis, h
and cancer.

Another theory that seeks to explain the Hayflick limit is related to
telomeres, the noncoding bits of DNA attached to the end of each
chromosome.[6] Telomeres act like the leaders in a filmstrip and ensure
that the DNA is accurately replicated. Cell division involves the split-
ting apart of the two strands of the DNA molecule and their reconstitution into complete new copies of the molecule in the daughter
cells. But with each cell division, the telomeres get a bit shorter, until
they are unable to protect the ends of the DNA strand and the cell,
recognizing the short telomeres as damaged DNA, ceases growth.
Dolly the sheep, cloned from somatic cells of an adult animal, had the
shortened telomeres of an adult rather than the longer ones of a newborn lamb, and presumably will not live as long as a naturally born
sibling.

There are three major types of cells that are not subject to the
Hayflick limit: germ cells, cancer cells, and certain types of stem
cells. The reason these cells can reproduce indefinitely has to do with
the presence of an enzyme called telomerase, first isolated in 1989,
which prevents the shortening of telomeres. This is what permits the
germ line to continue through the generations without end, and is
also what lies behind the explosive growth of cancer tumors.

Leonard Guarente of the Massachusetts Institute of Technology
reported findings that calorie restriction in yeast increased longevity,
through the action of a single gene known as SIR2 (silent information
regulator No. 2). The SIR2 gene represses genes that generate riboso-
mal wastes that build up in yeast cells and lead to their eventual
death; low-calorie diets restrict reproduction but are helpful to the
functioning of the SIR2 gene. This may provide a molecular explanation for why laboratory rats fed a low-calorie diet live up to 40 percent
longer than other rats.[7]

Biologists such as Guarente have suggested that there might
someday be a relatively simple genetic route to life extension in humans: while it is not practical to feed people such restricted diets,
there may be other ways of enhancing the functioning of the SIR

genes. Other gerontologists, such as Tom Kirkwood, assert flatly that aging is the result of a complex series of processes at the level of cells, organs, and the body as a whole, and that there is therefore no single, simple mechanism that controls aging and death.[8]

If a genetic shortcut to immortality exists, the race is already on within the biotech industry to find it. The Geron Corporation has already cloned and patented the human gene for telomerase and, along with Advanced Cell Technology, has an active research program into embryonic stem cells. The latter are cells that make up an embryo at the earliest stages of development, before there has been any differentiation into different types of tissue and organs. Stem cells have the potential to become any cell or tissue in the body, and hence hold the promise of generating entirely new body parts to replace ones worn out through the aging process. Unlike organs transplanted from donors, such cloned body parts will be almost genetically identical to cells in the body into which they are placed, and so presumably free from the kinds of immune reactions that lead to transplant rejection.

Stem cell research represents one of the great frontiers of contemporary biomedical research. It is also hugely controversial as a result of its use of embryos as sources of stem cells—embryos which must be destroyed in the process.[9] The embryos usually come from the extra embryos "banked" by in vitro fertilization clinics. (Once created, stem cell "lines" can be replicated almost indefinitely.) Out of concern that stem cell research would encourage abortion or lead to the deliberate destruction of human embryos, the U.S. Congress imposed a ban on funding from the National Institutes of Health for research that could harm embryos,[10] pushing U.S. stem cell research into the private sector. In 2001 a bitter policy debate exploded in the United States as the Bush administration considered lifting the ban. In the end, the administration decided to permit federally funded research, but only on the sixty or so existing stem cell lines that had already been created.

It is impossible to know at this point whether the biotech industry will eventually be able to come up with a shortcut to the prolongation of life, such as a simple pill that will add another decade or two to people's life spans.[11] Even if this never happens, however, it seems fairly safe to say that the *cumulative* impact of all the biomedical re-

search going on at present will be to further increase life exp
over time and therefore to continue the trend that has b
way for the last century. So it is not at all premature to think ...
some of the political scenarios and social consequences that might
emerge from demographic trends that are already well under way.

In Europe at the beginning of the eighteenth century, half of all
children died before they reached the age of 15. The French demographer Jean Fourastié has pointed out that reaching the age of 52 was
then an accomplishment, since only a small minority of the population did so, and that such a person might legitimately consider himself or herself a "survivor."[12] Since most people reached the peak of
their productive lives during their 40s and 50s, a huge amount of human potential was wasted. In the 1990s, by contrast, over 83 percent
of the population could expect to live to the age of 65, and more than
28 percent would still be alive at age 85.[13]

Increasing life expectancies are only part of the story of what has
happened to populations in the developed world by the end of the
twentieth century. The other major development has been the dramatic fall in fertility rates. Countries such as Italy, Spain, and Japan
have total fertility rates (that is, the average number of children born
to a woman in her lifetime) of between 1.1 and 1.5, far below the replacement rate of about 2.2. The combination of falling birthrates and
increasing life expectancies has dramatically shifted the age distribution in developed countries. While the median age of the U.S. population was about 19 years in 1850, it had risen to 34 years by the
1990s.[14] This is nothing compared to what will happen in the first half
of the twenty-first century. While the median age in the United States
will climb to almost 40 by the year 2050, the change will be even more
dramatic in Europe and Japan, where rates of immigration and fertility are lower. In the absence of an unanticipated increase in fertility,
the demographer Nicholas Eberstadt estimates, based on UN data,
that the median age in Germany will be 54, in Japan 56, and in Italy
58.[15] These estimates, it should be noted, do *not* assume any dramatic
increases in life expectancies. If only some of the promises of biotechnology for gerontology pan out, it could well be the case that *half* of
the populations of developed countries will be retirement age or older
by this point.

Up to now, the "graying" of the populations of developed countries has been discussed primarily in the context of the social security liability that it will create. This looming crisis is real enough: Japan, for instance, will go from a situation in which there were four active workers for every retired person at the end of the twentieth century, to one in which there are only two workers per retired person a generation or so down the road. But there are other political implications as well.

Take international relations.[16] While some developing countries have succeeded in approaching or even crossing the demographic transition to subreplacement fertility and declining population growth, as the developed world has, many of the poorer parts of the world, including the Middle East and sub-Saharan Africa, continue to experience high rates of growth. This means that the dividing line between the First and Third Worlds in two generations will be a matter not simply of income and culture but of age as well, with Europe, Japan, and parts of North America having a median age of nearly 60 and their less developed neighbors having median ages somewhere in the early 20s.

In addition, voting age populations in the developed world will be more heavily feminized, in part because more women in the growing elderly cohort will live to advanced ages than men, and in part because of a long-term sociological shift toward greater female political participation. Indeed, elderly women will emerge as one of the most important blocs of voters courted by twenty-first-century politicians.

What this will mean for international politics is of course far from clear, but we do know on the basis of past experience that there are important differences in attitudes toward foreign policy and national security between women and men, and between older and younger voters. American women, for example, have always been less supportive than American men of U.S. involvement in war, by an average margin of seven to nine percentage points. They are also consistently less supportive of defense spending and the use of force abroad. In a 1995 Roper survey conducted for the Chicago Council on Foreign Relations, men favored U.S. intervention in Korea in the event of a North Korean attack by a margin of 49 to 40 percent, while women were opposed by a margin of 30 to 54. Fifty-four percent of men felt that it was important to maintain superior worldwide military power, compared with only 45 percent of women. Women, moreover, are less likely than men to see force as a legitimate tool for resolving conflicts.[17]

Developed countries will face other obstacles to the use of force. Elderly people, and particularly elderly women, are not the first to be called to serve in military organizations, so the pool of available military manpower will shrink. The willingness of people in such societies to tolerate battle casualties among their young may fall as well.[18] Nicholas Eberstadt estimates that given current fertility trends, Italy in 2050 will be a society in which only 5 percent of all children have any collateral relatives (that is, brothers, sisters, aunts, uncles, cousins, and so forth) at all. People will be primarily related to their parents, grandparents, great-grandparents, and to their own offspring. Such a tenuous generational line is likely to increase the reluctance to go to war and accept death in battle.

The world may well be divided, then, between a North whose political tone is set by elderly women, and a South driven by what Thomas Friedman labels super-empowered angry young men. It was a group of such men that carried out the September 11 attacks on the World Trade Center. This does not, of course, mean that the North will fail to rise to challenges posed by the South, or that conflict between the two regions is inevitable. Biology is not destiny. But politicians will have to work within frameworks established by basic demographic facts, and one of those facts may be that many countries in the North will be both shrinking and aging.

There is another, perhaps more likely, scenario that will bring these worlds into direct contact: immigration. The estimates of falling populations in Europe and Japan given above assume no large increases in net immigration. This is unlikely, however, simply because developed countries will want economic growth and the population necessary to sustain it. This means that the North-South divide will be replicated within each country, with an increasingly elderly native-born population living alongside a culturally different and substantially younger immigrant population. The United States and other English-speaking countries have traditionally been good at assimilating culturally diverse groups of immigrants, but other countries, such as Germany and Japan, have not. Europe has already seen the rise of anti-immigrant backlash movements, such as the National Front in

France, the Vlaams Blok in Belgium, the Lega Lombarda in Italy, and Jörg Haider's Freedom Party in Austria. For these countries, changes in the age structure of their populations, abetted by increasing longevity, are likely to lay the ground for growing social conflict.

The prolongation of life through biotechnology will have dramatic effects on the internal structures of societies as well. The most important of these has to do with the management of social hierarchies.

Human beings are by nature status-conscious animals who, like their primate cousins, tend from an early age to arrange themselves in a bewildering variety of dominance hierarchies.[19] This hierarchical behavior is innate and has easily survived the arrival of modern ideologies like democracy and socialism that purport to be based on universal equality. (One has only to look at pictures of the politburos of the former Soviet Union and China, where the top leadership is arrayed in careful order of dominance.) The nature of these hierarchies has changed as a result of cultural evolution, from traditional ones based on physical prowess or inherited social status, to modern ones based on cognitive ability or education. But their hierarchical nature remains.

If one looks around at a society, one quickly discovers that many of these hierarchies are age-graded. Sixth graders feel themselves superior to fifth graders and dominate the playground if both have recess together; tenured professors lord it over untenured ones and carefully control entry into their august circle. Age-graded hierarchies make functional sense insofar as age is correlated in many societies with physical prowess, learning, experience, judgment, achievement, and the like. But past a certain age, the correlation between age and ability begins to go in the opposite direction. With life expectancies only in the 40s or 50s for most of human history, societies could rely on normal generational succession to take care of this problem. Mandatory retirement ages came into vogue only in the late nineteenth century, when increasing numbers of people began to survive into old age.*

Life extension will wreak havoc with most existing age-graded hi-

---

*Bismarck, who established Europe's first social security system, set retirement at 65, an age to which virtually no one at that time lived.

erarchies. Such hierarchies traditionally assume a pyramidal structure because death winnows the pool of competitors for the top ranks, abetted by artificial constraints such as the widely held belief that everyone has the "right" to retire at age 65. With people routinely living and working into their 60s, 70s, 80s, and even 90s, however, these pyramids will increasingly resemble squat trapezoids or even rectangles. The natural tendency of one generation to get out of the way of the up-and-coming one will be replaced by the simultaneous existence of three, four, even five generations.

We have already seen the deleterious consequences of prolonged generational succession in authoritarian regimes that have no constitutional requirements limiting tenure in office. As long as dictators like Francisco Franco, Kim Il Sung, and Fidel Castro physically survive, their societies have no way of replacing them, and all political and social change is effectively on hold until they die.[20] In the future, with technologically enhanced life spans, such societies may find themselves locked in a ludicrous deathwatch not for years but for decades.

In societies that are more democratic and/or meritocratic, there are institutional mechanisms for removing leaders, bosses, or CEOs who are past their prime. But the problem does not go away by any stretch of the imagination.

The root problem lies, of course, in the fact that people at the top of social hierarchies generally do not want to lose status or power and will often use their considerable influence to protect their positions. Age-related declines in capabilities have to be fairly pronounced before other people will go to the trouble of removing a leader, boss, ballplayer, professor, or board member. Impersonal formal rules like mandatory retirement ages are useful precisely because they don't require institutions to make nuanced personal judgments about an individual older person's capability. But impersonal rules often discriminate against older people who are perfectly capable of continuing to work and for that reason have been abolished in many American workplaces.

There is at present a tremendous amount of political correctness regarding age: *ageism* has entered the pantheon of proscribed prejudices, next to racism, sexism, and homophobia. There is of course discrimination against older people, particularly in a youth-obsessed society like that of the United States. But there are also a number of reasons why generational succession is a good thing. Chief among them is that it is a major stimulant of progress and change.

Many observers have noted that political change often occurs at generational intervals—from the Progressive Era to the New Deal, from the Kennedy years to Reaganism.[21] There is no mystery as to why this is so: people born in the same age cohort experience major life events—the Great Depression, World War II, or the sexual revolution—together. Once people's life views and preferences have been formed by these experiences, they may adapt to new circumstances in small ways, but it is very difficult to get them to change broad outlooks. A black person who grew up in the old South has a hard time seeing a white cop as anything but an untrustworthy agent of an oppressive system of racial segregation, regardless of whether this makes sense given the realities of life in a northern city. Those who lived through the Great Depression cannot help feeling uneasy at the lavish spending habits of their grandchildren.

This is true not just in political but in intellectual life as well. There is a saying that the discipline of economics makes progress one funeral at a time, which is unfortunately truer than most people are willing to admit. The survival of a basic "paradigm" (for example, Keynesianism or Friedmanism) that shapes the way most scientists and intellectuals think about things at a particular time depends not just on empirical evidence, as some would like to think, but on the physical survival of the people who created that paradigm. As long as they sit on top of age-graded hierarchies like peer review boards, tenure committees, and foundation boards of trustees, the basic paradigm will often remain virtually unshakable.

It stands to reason, then, that political, social, and intellectual change will occur much more slowly in societies with substantially longer average life spans. With three or more generations active and working at the same time, the younger age cohorts will never constitute more than a small minority of voices clamoring to be heard, and generational change will never be fully decisive. To adjust more rapidly, such societies will have to establish rules mandating constant retraining and downward social mobility at later stages in life. The idea that one can acquire skills and education during one's 20s that will remain useful for the next forty years is implausible enough at present, given the pace of technological change. The idea that these skills would remain relevant over working lives of fifty, sixty, or seventy years becomes even more preposterous. Older people will have to move down the social hierarchy not just to retrain but to make room for new entrants coming up from the bottom. If they don't, generational warfare will join class and ethnic conflict as a major dividing line in society. Getting older people out of the way of younger ones will become a significant struggle, and societies may have to resort to impersonal, institutionalized forms of ageism in a future world of expanded life expectancies.

Other social effects of life extension will depend heavily on the exact way that the geriatric revolution plays itself out—that is, whether people will remain physically and mentally vigorous throughout these lengthening life spans, or whether society will increasingly come to resemble a giant nursing home.

The medical profession is dedicated to the proposition that anything that can defeat disease and prolong life is unequivocally a good thing. The fear of death is one of the deepest and most abiding human passions, so it is understandable that we should celebrate any advance in medical technology that appears to put death off. But people worry about the quality of their lives as well—not just the quantity. Ideally, one would like not merely to live longer but also to have one's different faculties fail as close as possible to when death finally comes, so that one does not have to pass through a period of debility at the end of life.

While many medical advances have increased the quality of life for older people, many have had the opposite effect by prolonging only one aspect of life and increasing dependency. Alzheimer's disease—in which certain parts of the brain waste away, leading to loss of memory and eventually dementia—is a good example of this, because the likelihood of getting it rises proportionately with age. At age 65, only one person in a hundred is likely to come down with Alzheimer's; at 85, it is one in six.[22] The rapid growth in the population suffering from Alzheimer's in developed countries is thus a direct result of increased life expectancies, which have prolonged the health of the body without prolonging resistance to this terrible neurological disease.

There are in fact two periods of old age that medical technology has opened up, at least for people in the developed world.[23] Category I extends from age 65 until sometime in one's 80s, when people can increasingly expect to live healthy and active lives, with enough resources to take advantage of them. Much of the happy talk about increased longevity concerns this period, and indeed the emergence of this new phase of life as a realistic expectation for most people is an achievement of which modern medicine can be proud. The chief problem for people in this category will be the encroachment of working life on their domain: for simple economic reasons, there will be powerful pressures to raise retirement ages and keep the over-65 cohort in the workforce for as long as possible. This does not imply any kind of social disaster: older workers may have to retrain and accept some degree of downward social mobility, but many of them will welcome the opportunity to contribute their labor to society.

The second phase of old age, Category II, is much more problematic. It is the period that most people currently reach by their 80s, when their capabilities decline and they return increasingly to a child-like state of dependency. This is the period that society doesn't like to think about, much less experience, since it flies in the face of ideals of personal autonomy that most people hold dear. Increases in the number of people in both Category I and Category II have created a novel situation in which individuals approaching retirement age today find their own choices constrained by the fact that they still have an elderly parent alive and dependent on them for care.

The social impact of ever-increasing life expectancies will depend on the relative sizes of these two groups, which in turn will depend on the "evenness" of future life-prolonging advances. The best scenario would be one in which technology simultaneously pushes back parallel aging processes—for instance, by the discovery of a common molecular source of aging in all somatic cells, and the delaying of this process throughout the body. Failure of the different parts would come at the same time, just later; people in Category I would be more numerous and those in Category II less so. The worst scenario would be one of highly uneven advance, in which, for example, we found ways to preserve bodily health but could not put off age-related mental deterioration. Stem cell research might yield ways to grow new body parts, as William Haseltine is quoted as suggesting at the beginning of Chapter 2. But without a parallel cure for Alzheimer's disease, this wonderful new technology would do no more than allow more people to persist in vegetative states for years longer than is currently possible.

An explosion in the number of people in Category II might be labeled the national nursing home scenario, in which people routinely live to be 150 but spend the last fifty years in a state of childlike dependence on caretakers. There is of course no way of predicting whether this or the happier extension of the Category I period will play itself out. If there is no molecular shortcut to postponing death because aging is the result of the gradual accumulation of damage to a wide range of different biological systems, then there is no reason to think that future medical advances will proceed with a neat simultaneity, any more than they have in the past. That existing medical technology is capable only of keeping people's bodies alive at a much reduced quality of life is the reason assisted suicide and euthanasia, as well as figures like Jack Kevorkian, have come to the fore as public issues in the United States and elsewhere in recent years.

In the future, biotechnology is likely to offer us bargains that trade off length of life span for quality of life. If they are accepted, the social consequences could be dramatic. But assessing them will be very difficult: slight changes in mental capabilities such as loss of short-term memory or growing rigidity in one's beliefs are inherently difficult to measure and evaluate. The political correctness about aging noted earlier will make a truly frank assessment nearly impossible, both for individuals dealing with elderly relatives and for societies trying to formulate public policies. To avoid any hint of discrimination against older people, or the suggestion that their lives are somehow worth less than those of the young, anyone who writes on the future of aging feels compelled to be relentlessly sunny in predicting that medical advances will increase both the quantity and quality of life.

This is most evident with regard to sexuality. According to one writer on aging, "One of the factors inhibiting sexuality with ageing is undoubtedly the brain-washing that all of us experience which says that the older person is less sexually attractive."[24] Would that sexuality were only a matter of brainwashing! Unfortunately, there are good Darwinian reasons that sexual attractiveness is linked to youth, particularly in women. Evolution has created sexual desire for the purpose of fostering reproduction, and there are few selective pressures for humans to develop sexual attraction to partners past their prime reproductive years.[25] The consequence is that in another fifty years, most developed societies may have become "postsexual," in the sense that the vast majority of their members will no longer put sex at the top of their "to do" lists.

There are a number of unanswerable questions about what life in this kind of future would be like, since there have never in human history been societies with median ages of 60, 70, or higher. What would such a society's self-image be? If you go to a typical airport newsstand and look at the people pictured on magazine covers, their median age is likely to be in the low 20s, the vast majority good-looking and in perfect health. For most historical human societies, these covers would have reflected the actual median age, though not the looks or health, of the society as a whole. What will magazine covers look like in another couple of generations, when people in their early 20s constitute only a tiny minority of the population? Will society still want to think of itself as young, dynamic, sexy, and healthy, even though the image departs from the reality that people see around them to an even more extreme degree than today? Or will tastes and habits shift, with the youth culture going into terminal decline?

A shift in the demographic balance toward societies with a majority of people in Categories I and II will have much more profound implications for the meaning of life and death as well. For virtually all of human history up to the present, people's lives and identities were bound up either with reproduction—that is, having families and raising children—or with earning the resources to support themselves and their families. Family and work both enmesh individuals in a web of social obligations over which they frequently have little control and which are a source of struggle and anxiety but also of tremendous satisfaction. Learning to meet those social obligations is a source of both morality and character. People in Categories I and II, by contrast, will have a much more attenuated relationship to both family and work.

They will be beyond reproductive years, with links primarily to ancestors and descendants. Some in Category I may choose to work, but the obligation to work and the kinds of mandatory social ties that work engenders will be replaced largely by a host of elective occupations. Those in Category II will not reproduce, not work, and indeed will see a flow of resources and obligation moving one way: toward them.

This does not mean that people in either category will suddenly become irresponsible or footloose. It does mean, however, that they may find their lives both emptier and lonelier, since it is precisely those obligatory ties that make life worth living for many people. When retirement is seen as a brief period of leisure following a life of hard work and struggle, it may seem like a well-deserved reward; if it stretches on for twenty or thirty years with no apparent end, it may seem simply pointless. And it is hard to see how a prolonged period of dependency or debility for people in Category II will be experienced as joyful or fulfilling.

People's relationship to death will change as well. Death may come to be seen not as a natural and inevitable aspect of life, but a preventable evil like polio or the measles. If so, then accepting death will appear to be a foolish choice, not something to be faced with dignity or nobility. Will people still be willing to sacrifice their lives for others, when their lives could potentially stretch out ahead of them indefinitely, or condone the sacrifice of the lives of others? Will they cling desperately to the life that biotechnology offers? Or might the prospect of an unendingly empty life appear simply unbearable?

## GENETIC ENGINEERING

**"All beings so far have created something beyond themselves; and do you
want to be the ebb of this great flood and even go back to the beasts
rather than overcome man? What is the ape to man? A laughingstock or a
painful embarrassment. And man shall be just that for the overman: a
laughingstock or a painful embarrassment. You have made your way from
worm to man, and much in you is still worm. Once you were apes, and
even now, too, man is more ape than any ape."**

**Friedrich Nietzsche, *Thus Spoke Zarathustra* I.3**

All of the consequences described in the preceding three chapters may come to pass without any further progress in the most revolutionary biotechnology of all, genetic engineering. Today, genetic engineering is used commonly in agricultural biotechnology to produce genetically modified organisms such as Bt corn (which produces its own insecticide) or Roundup Ready soybeans (which are resistant to certain weed-control herbicides), products that have been the focus of controversy and protest around the world. The next line of advance is obviously to apply this technology to human beings. Human genetic engineering raises most directly the prospect of a new kind of eugenics, with all the moral implications with which that word is fraught, and ultimately the ability to change human nature.

Yet despite completion of the Human Genome Project, contemporary biotechnology is today very far from being able to modify human DNA in the way that it can modify the DNA of corn or beef cattle. Some people would argue that we will never in fact achieve this kind of capability and that the ultimate prospects for genetic technology have been grossly overhyped both by ambitious scientists and by biotechnology companies out for quick profits. Changing human nature is neither possible, according to some, nor remotely on the agenda of contemporary biotechnology. We need, then, a balanced assessment of what this technology can be expected to achieve, and a sense of the constraints that it may eventually face.

The Human Genome Project was a massive effort, funded by the United States and other governments, to decode the entire DNA sequence of a human being, just as the DNA sequences of lesser creatures, like nematodes and yeast, had been decoded.[1] DNA molecules are the famous twisted, double-stranded sequences of four bases that make up each of the forty-six chromosomes contained in the nucleus of every cell in the body. These sequences constitute a digital code that is used to synthesize amino acids, which are then combined to produce the proteins that are the building blocks of all organisms. The human genome consists of some 3 billion pairs of bases, a large percentage of which consists of noncoding, "silent" DNA. The remainder constitutes genes that contain the actual blueprints for human life.*

The complete sequencing of the human genome was completed way ahead of schedule, in June 2000, in part because of competition between the official government-sponsored Human Genome Project and a similar effort by a private biotech company, Celera Genomics. The publicity surrounding this event sometimes suggested that scientists had decoded the genetic basis of life, but all the sequencing did was present the transcript of a book written in a language that is only partially understood. There is great uncertainty on such basic issues as how many genes are contained in human DNA. A few months af-

---

*Those who are interested in seeing exactly what the raw code looks like, and how each chromosome is divided into genes and noncoding areas, can simply look at the Web site of the National Institutes of Health's National Center for Biotechnology Information at http://www.ncbi.nlm.nih.gov/Genbank/GenbankOverview.html.

ter completion of the sequencing, Celera and the International Human Genome Sequencing Consortium released a study indicating that the number was 30,000 to 40,000 instead of the more than 100,000 previously estimated. Beyond genomics lies the burgeoning field of proteomics, which seeks to understand how genes code for proteins and how the proteins themselves fold into the exquisitely complex shapes required by cells.[2] And beyond proteomics there lies the unbelievably complex task of understanding how these molecules develop into tissues, organs, and complete human beings.

The Human Genome Project would not have been possible without parallel advances in the information technology required to record, catalog, search, and analyze the billions of bases making up human DNA. The merger of biology and information technology has led to the emergence of a new field, known as bioinformatics.[3] What will be possible in the future will depend heavily on the ability of computers to interpret the mind-boggling amounts of data generated by genomics and proteomics and to build reliable models of phenomena such as protein folding.

The simple identification of genes in the genome does not mean that anyone knows what it is they do. A great deal of progress has been made in the past two decades in finding the genes connected to cystic fibrosis, sickle-cell anemia, Huntington's chorea, Tay-Sachs disease, and the like. But these have all tended to be relatively simple disorders, in which the pathology can be traced to a wrong allele, or coding sequence, in a single gene. Other diseases are caused by multiple genes that interact in complex ways: some genes control the expression (that is, the activation) of other genes, some interact with the environment in complex ways, some produce two or more effects, and some produce effects that will not be visible until late in the organism's life cycle.

When it comes to higher-order conditions and behaviors, such as intelligence, aggression, sexuality, and the like, we know nothing more today than that there is some degree of genetic causation, from studies in behavior genetics. We have no idea what genes are ultimately responsible, but suspect that the causal relationships are extraordinarily complex. In the words of Stuart Kauffman, founder and chief scientific officer of BiosGroup, these genes are "some kind of

parallel-processing chemical computer in which genes ar
ously turning one another on and off in some vastly compl
of interaction. Cell-signaling pathways are linked to genetic
pathways in ways we're just beginning to unscramble."[4]

The first step toward giving parents greater control over the genetic makeup of their children will come not from genetic engineering but with preimplantation genetic diagnosis and screening. In the
future it should be routinely possible for parents to have their embryos automatically screened for a wide variety of disorders, and those
with the "right" genes implanted in the mother's womb. Present-day
medical technology, such as amniocentesis and sonograms, gives parents a certain degree of choice already, as when a fetus diagnosed
with Down's syndrome is aborted, or when girl fetuses are aborted in
Asia. Embryos have already been successfully screened for birth defects like cystic fibrosis.[5] Geneticist Lee Silver paints a future
scenario in which a woman produces a hundred or so embryos, has
them automatically analyzed for a "genetic profile," and then with a
few clicks of the mouse selects the one that not only lacks alleles for
single-gene disorders like cystic fibrosis, but also has enhanced characteristics, such as height, hair color, and intelligence.[6] The technologies to bring this about do not exist now but are on the way: a
company called Affymetrix, for example, has developed a so-called
DNA chip that automatically screens a DNA sample for various
markers of cancer and other disorders.[7] Preimplantation diagnosis
and screening does not require any ability to manipulate the embryo's
DNA, but limits parental choice to the kind of variation that normally
occurs through sexual reproduction.

The other technology that is likely to mature well before human
genetic engineering is human cloning. Ian Wilmut's success in creating the cloned sheep Dolly in 1997 provoked a huge amount of controversy and speculation about the possibility of cloning a human
being from adult cells.[8] President Clinton's request to the National
Bioethics Advisory Commission for advice on this subject led to a
study that recommended a ban on federal funding for human cloning
research, a moratorium on such activities by private companies and
concerns, and consideration by Congress of a legislative ban.[9] In lieu
of a congressional ban, however, the attempt to clone a human being

by a non–federally funded organization remains legal. There are reports that a sect called the Raelians is trying to do just that,[10] as well as a well-publicized effort by Severino Antinori and Panos Zavos. The technical obstacles to human cloning are substantially smaller than in the case of either preimplantation diagnosis or genetic engineering, and have mostly to do with the safety and ethicality of experimenting with human beings.

## THE ROAD TO DESIGNER BABIES

The ultimate prize of modern genetic technology will be the "designer baby."[11] That is, geneticists will identify the "gene for" a characteristic like intelligence, height, hair color, aggression, or self-esteem and use this knowledge to create a "better" version of the child. The gene in question may not even have to come from a human being. This is, after all, what happens in agricultural biotechnology. Bt corn, first developed by Ciba Seeds (now Novartis Seeds) and Mycogen Seeds in 1996, has an exotic gene inserted into its DNA that allows it to produce a protein from the *Bacillus thuringiensis* bacterium (hence the Bt designation) that is toxic to insect pests such as the European corn borer. The resulting plant is thus genetically modified to produce its own pesticide, and it hands down this characteristic to its offspring.

Doing the same thing to human beings is, of all of the technologies discussed in this chapter, the most remote. There are two ways by which genetic engineering can be accomplished: somatic gene therapy and germ-line engineering. The first attempts to change the DNA within a large number of target cells, usually by delivering the new, modified genetic material by means of a virus or "vector." A number of somatic gene therapy trials have been conducted in recent years, with relatively little success. The problem with this approach is that the body is made up of trillions of cells; for the therapy to be effective, the genetic material of what amounts to millions of cells has to be altered. The somatic cells in question die with the individual being treated, if not before; the therapy has no lingering generational effects.

Germ-line engineering, by contrast, is what is done routinely in agricultural biotechnology and has been successfully carried out in a wide variety of animals. Modification of the germ line requires, at least in theory, changing only one set of DNA molecules, those in the fertilized egg, which will eventually undergo division and ramify into a complete human being. While somatic gene therapy changes only the DNA of somatic cells, and therefore affects only the individual who receives the treatment, germ-line changes are passed down to the individual's offspring. This has obvious attractions for the treatment of inherited diseases, such as diabetes.[12]

Among other new technologies currently under study are artificial chromosomes that would add an extra chromosome to the forty-six natural ones; the chromosone could be turned on only when the recipient was old enough to give his or her informed consent and would not be inherited by descendants.[13] This technique would avoid the need to alter or replace genes in existing chromosomes. Artificial chromosomes might thus constitute a bridge between preimplantation screening and permanent modification of the germ line.

Before human beings can be genetically modified in this manner, however, a number of steep obstacles need to be overcome. The first has to do with the sheer complexity of the problem, which suggests to some that any meaningful kind of genetic engineering for higher-order behaviors will simply be impossible. We noted earlier that many diseases are caused by the interaction of multiple genes; it is also the case that a single gene has multiple effects. It was believed at one time that each gene produced one messenger RNA, which in turn produced one protein. But if the human genome in fact contains closer to 30,000 than 100,000 genes, then this model cannot hold up, since there are far more than 30,000 proteins making up the human body. This suggests that single genes play a role in producing many proteins and therefore have multiple functions. The allele responsible for sickle-cell anemia, for example, also confers resistance to malaria, which is why it is common among blacks, who trace their ancestry to Africa, where malaria was a major disease. Repairing the gene for sickle-cell anemia might therefore increase susceptibility to malaria, something that may not matter much for people in North America but would harm carriers of the new gene in Africa. Genes have been compared to an ecosystem, where each part influences every other part: in the words of Edward O. Wilson, "in heredity as in the environment, you cannot do just one thing. When a gene is changed by mutation or replaced by another gene, unexpected and possibly unpleasant side effects are likely to follow."[14]

The second major obstacle to human genetic engineering has to do with the ethics of human experimentation. The National Bioethics Advisory Commission raised the danger of human experimentation as the chief reason for seeking a short-term ban on human cloning. It took nearly 270 failed attempts before Dolly was successfully cloned.[15] While many of these failures came at the implantation stage, nearly 30 percent of all animals that have been cloned since then have been born with serious abnormalities. As noted earlier, Dolly was born with shortened telomeres and will probably not live as long as a sheep born normally. One would presumably not want to create a human baby until one had a much higher chance of success, and even then the cloning process might produce defects that wouldn't show up for years.

The dangers that exist for cloning would be greatly magnified in the case of genetic engineering, given the multiple causal pathways between genes and their ultimate expression in the phenotype.[16] The Law of Unintended Consequences would apply here in spades: a gene affecting one particular disease susceptibility might have secondary or tertiary consequences that are unrecognized at the time that the gene is reengineered, only to show up years or even a generation later.

The final constraint on any future ability to modify human nature has to do with populations. Even if human genetic engineering overcomes these first two obstacles (that is, complex causality and the dangers of human experimentation) and produces a successful designer baby, "human nature" will not be altered unless such changes occur in a statistically significant way for the population as a whole. The Council of Europe has recommended the banning of germ-line engineering on the grounds that it would affect the "genetic patrimony of mankind." This particular concern, as a number of critics have pointed out, is a bit silly: the "genetic patrimony of mankind" is a very large gene pool containing many different alleles. Modifying, eliminating, or adding to those alleles on a small scale will change an individual's patrimony but not the human race's. A handful of rich people genetically modifying their children for greater height or intelligence would have no effect on species-typical height or IQ. Fred Iklé argues that any future attempt to eugenically improve the human race would be quickly overwhelmed by natural population growth.[17]

Do these constraints on genetic engineering, then, mean that any meaningful alteration of human nature is off the table for the foreseeable future? There are several reasons to be cautious in coming to such a judgment prematurely.

The first has to do with the remarkable and largely unanticipated speed of scientific and technological developments in the life sciences. In the late 1980s there was a firm consensus among geneticists that it was impossible to clone a mammal from adult somatic cells, a view that came to an end with Dolly in 1997.[18] As recently as the mid-1990s, geneticists were predicting that the Human Genome Project would be completed sometime between 2010 and 2020; the actual date by which the new, highly automated sequencing machines completed the work was July 2000. There is no way of predicting what kinds of shortcuts may appear in future years to reduce the complexity of the task ahead. For example, the brain is the archetype of a so-called complex adaptive system—that is, a system made up of numerous agents (in this case, neurons and other brain cells) following relatively simple rules that produce highly complex emergent behavior at a system level. Any attempt to model a brain using brute-force computation methods—one which tries to duplicate all of the billions of neuronal connections—is almost certainly bound to fail. A complex adaptive model, on the other hand, that seeks to model system-level complexity as an emergent property might have a much greater chance of succeeding. The same may be true for the interaction of genes.

That the multiple functions of genes and gene interactions are highly complex does not mean that all human genetic engineering will be on hold until we fully understand them. No technology ever develops in this fashion. New drugs are invented, tested, and approved for use all the time without the manufacturers knowing exactly how they produce their effects. It is often the case in pharmacology that side effects go unrecognized, sometimes for years, or that a drug will interact with other drugs or conditions in ways that were totally unanticipated when it was first introduced. Genetic engineers will tackle simple problems first, and then work their way up the ladder of complexity. While it is likely that higher-order behaviors are the result of the complex interactions of many genes, we don't know that this is invariably the case. We may stumble on relatively simple genetic interventions that produce dramatic changes in behavior.

The issue of human experimentation is a serious obstacle to rapid development of genetic engineering but by no means an insuperable one. As in drug testing, animals will bear most of the burden of risk at first. The kinds of risks acceptable in human trials will depend on projected benefits: a disease like Huntington's chorea, which produces a one-in-two chance of dementia and death in individuals and their offspring who carry the wrong allele, will be treated differently from an enhancement of muscle tone or breast size. The mere fact that there may be unanticipated or long-term side effects will not deter people from pursuing genetic remedies, any more than it has in earlier phases of medical development.

The question of whether the eugenic or dysgenic effects of genetic engineering could ever become sufficiently widespread to affect human nature itself is similarly an open one. Obviously, any form of genetic engineering that could have significant effects on populations would have to be shown to be desirable, safe, and relatively cheap. Designer babies will be expensive at first and an option only for the well-to-do. Whether having a designer baby will ever become cheap and relatively popular will depend on how rapidly technologies like preimplantation diagnosis come down the cost curve.

There are precedents, however, for new medical technologies having population-level effects as a result of millions of individual choices. One has to look no further than contemporary Asia, where a combination of cheap sonograms and easy access to abortion has led to a dramatic shifting of sex ratios. In Korea, for example, 122 boys were born in the early 1990s for every 100 girls, compared with a normal ratio of 105 to 100. The ratio in the People's Republic of China is only somewhat lower, at 117 boys for every 100 girls, and there are parts of northern India where ratios are even more skewed.[19] This has led to a deficit of girls in Asia that the economist Amartya Sen at one point estimated to be 100 million.[20] In all of these societies, abortion for the purpose of sex selection is illegal; but despite government pressure, the desire of individual parents for a male heir has produced grossly lopsided sex ratios.

Highly skewed sex ratios can produce important social consequences. By the second decade of the twenty-first century, China will face a situation in which up to one fifth of its marriage-age male population will not be able to find brides. It is hard to imagine a better formula for trouble, given the propensity of unattached young males to be involved in activities like risk-taking, rebellion, and crime.[21] There will be compensating benefits as well: the deficit of women will allow females to control the mating process more effectively, leading to more stable family life for those who can get married.*

Nobody knows whether genetic engineering will one day become as cheap and accessible as sonograms and abortion. Much depends on what its benefits are assumed to be. The most common fear expressed by present-day bioethicists is that only the wealthy will have access to this kind of genetic technology. But if a biotechnology of the future produces, for example, a safe and effective way to genetically engineer more intelligent children, then the stakes would immediately be raised. Under this scenario it is entirely plausible that an advanced, democratic welfare state would reenter the eugenics game, intervening this time not to prevent low-IQ people from breeding, but to help genetically disadvantaged people raise their IQs and the IQs of their offspring.[22] It would be the state, under these circumstances, that would make sure that the technology became cheap and accessible to all. And at that point, a population-level effect would very likely emerge.

That human genetic engineering will lead to unintended consequences and that it may never produce the kinds of effects some people hope for are not arguments that it will never be attempted. The

---

*Marcia Guttentag and Paul Secord have suggested that the sexual revolution and the breakdown of the traditional family in the United States was produced in part by sex ratios favoring men in the 1960s and 1970s. See Marcia Guttentag and Paul F. Secord, *Too Many Women? The Sex Ratio Question* (Newbury Park, Calif.: Sage Publications, 1983).

history of technological development is littered with new technologies that produced long-term consequences that led to their modification or even abandonment. For instance, no large hydroelectric projects have been undertaken anywhere in the developed world for the past couple of generations, despite periodic energy crises and rapidly growing demand for power.* The reason is that since the burst of dam building that produced the Hetch Hetchy Dam in 1923 and the Tennessee Valley Authority in the 1930s, an environmental consciousness has arisen that began to weigh the long-term environmental costs of hydroelectric power. When viewed today, the quasi-Stalinist movies that were made celebrating the heroic construction of Hoover Dam seem quaint in their glorification of the human conquest of nature and their blithe disregard of ecological consequences.

Human genetic engineering is only the fourth pathway to the future, and the most far-off stage in the development of biotechnology. We do not today have the ability to modify human nature in any significant way, and it may turn out that the human race will never achieve this ability. But two points need to be made.

First, even if genetic engineering never materializes, the first three stages of development in biotechnology—greater knowledge about genetic causation, neuropharmacology, and the prolongation of life —will all have important consequences for the politics of the twenty-first century. These developments will be hugely controversial because they will challenge dearly held notions of human equality and the capacity for moral choice; they will give societies new techniques for controlling the behavior of their citizens; they will change our understanding of human personality and identity; they will upend existing social hierarchies and affect the rate of intellectual, material, and political progress; and they will affect the nature of global politics.

The second point is that even if genetic engineering on a species level remains twenty-five, fifty, or one hundred years away, it is by far

*There have been major new hydroelectric projects, such as the Three Gorges Dam in China and the Ilisu Dam in Turkey, both of which have produced strong opposition from developed countries for their likely effects on the environment and on the populations in the floodplain, and, in the case of the Turkish dam, for the antiquities that will be covered by the floodwaters.

the most consequential of all future developments in biotechnology. The reason for this is that human nature is fundamental to our notions of justice, morality, and the good life, and all of these will undergo change if this technology becomes widespread. Why this is so will be taken up in Part II.

**6**

## WHY WE SHOULD WORRY

"Take Ectogenesis. Pfitzner and Kawaguchi had got the whole technique worked out. But would the Governments look at it? No. There was something called Christianity. Women were forced to go on being viviparous."

Aldous Huxley, *Brave New World*

In light of the possible pathways to the future laid out in the previous chapters, we need to ask the question: Why should we worry about biotechnology? Some critics, like the activist Jeremy Rifkin[1] and many European environmentalists, have been opposed to innovation in biotechnology virtually across the board. Given the very real medical benefits that will result from projected advances in human biotechnology, as well as the greater productivity and reduced use of pesticides coming from agricultural biotech, such categorical opposition is very difficult to justify. Biotechnology presents us with a special moral dilemma, because any reservations we may have about progress need to be tempered with a recognition of its undisputed promise.

Hanging over the entire field of genetics has been the specter of eugenics—that is, the deliberate breeding of people for certain selected heritable traits. The term *eugenics* was coined by Charles Darwin's cousin Francis Galton. In the late nineteenth and early twentieth centuries, state-sponsored eugenics programs attracted surprisingly broad support, not just from right-wing racists and social Darwinists, but from such progressives as the Fabian socialists Beatrice and Sidney Webb and George Bernard Shaw, the communists J.B.S. Haldane and J. D. Bernal, and the feminist and birth-control proponent Margaret Sanger.[2] The United States and other Western countries passed eugenics laws permitting the state to involuntarily sterilize people deemed "imbeciles," while encouraging people with desirable characteristics to have as many children as possible. In the words of Justice Oliver Wendell Holmes, "We want people who are healthy, good-natured, emotionally stable, sympathetic, and smart. We do not want idiots, imbeciles, paupers, and criminals."[3]

The eugenics movement in the United States was effectively terminated with revelations about the Nazis' eugenics policies, which involved the extermination of entire categories of people[4] and medical experimentation on people regarded as genetically inferior.[5] Since then, continental Europe has been effectively inoculated against any revival of eugenics and has, in fact, become inhospitable terrain for many forms of genetic research. The reaction against eugenics has not been universal: in progressive, social democratic Scandinavia, eugenics laws remained in effect until the 1960s.[6] Despite the fact that the Japanese conducted medical "experiments" on unwilling subjects during the Pacific War (through the activities of the infamous Unit 731), there has been a much smaller backlash against eugenics there and in most other Asian societies. China has pursued eugenics actively through its one-child population control policy and through a crude eugenics law, passed in 1995 and reminiscent of Western ones from the early twentieth century, that seeks to limit the right of low-IQ people to reproduce.[7]

There were two important objections to those earlier eugenics policies that would most likely not apply to any eugenics of the future, at least in the West.[8] The first was that eugenics programs could not achieve the ends they sought given the technology available at the time. Many of the defects and abnormalities against which the eugenicists thought they were selecting through forced sterilizations were the product of recessive genes—that is, genes that had to be inherited from both parents before they could be expressed. Many seemingly normal people would remain carriers of these genes and propagate those characteristics in the gene pool unless they could somehow be identified and sterilized as well. Many other "defects" were either not defects at all (for example, certain forms of low intelligence) or else were the result of nongenetic factors that could be remedied through better public health. For instance, certain villages in China have large populations of low-IQ children as a result not of bad heredity but of low levels of iodine in the children's diets.[9]

The second major objection to historical forms of eugenics is that they were state-sponsored and coercive. The Nazis, of course, carried this to horrifying extremes by killing or experimenting on "less desirable" people. But even in the United States it was possible for a court to decide that a particular individual was an imbecile or a moron (terms that were defined, as many mental conditions tend to be, very loosely) and to order that he or she be involuntarily sterilized. Given the view at the time that a wide variety of behaviors, such as alcoholism and criminality, were heritable, this gave the state potential dominion over the reproductive choices of a large part of its population. For observers like science writer Matt Ridley, state sponsorship is *the* primary problem with past eugenics laws; eugenics freely pursued by individuals has no similar stigma.[10]

Genetic engineering puts eugenics squarely back on the table, but it is clear that any future approach to eugenics will be very different from the historical varieties, at least in the developed West. The reason is that neither of these two objections is likely to apply, leading to the possibility of a kinder, gentler eugenics that will rob the word of some of the horror traditionally associated with it.

The first objection, that eugenics is not technically feasible, applies only to the kinds of technologies available in the early twentieth century, like forced sterilization. Advances in genetic screening currently allow doctors to identify carriers of recessive traits before they decide to have children, and in the future might allow them to identify embryos that carry a high risk of abnormality because they have inherited two recessive genes. Information of this sort is already available, for example, to individuals from a population such as Ashkenazi Jews, who have higher than normal probabilities of carrying the recessive Tay-Sachs gene; two such carriers may decide not to marry or to have children. In the future, germ-line engineering offers the possibility that such recessive genes could be eliminated from all subsequent descendants of a particular carrier. If the treatment were to become cheap and easy enough, it is possible to conceive of a particular gene being largely eliminated from entire populations.

The second objection to eugenics, that it was state-sponsored, is not likely to carry much weight in the future, because few modern societies are likely to want to get back into the eugenics game. Virtually all Western countries have moved sharply in the direction of stronger protection of individual rights since World War II, and the right to autonomy in reproductive decisions ranks high among those rights. The idea that states should legitimately worry about collective goods like the health of their national gene pools is no longer taken seriously but rather associated with outdated racist and elitist attitudes.

The kinder, gentler eugenics that is just over the horizon will then be a matter of individual choice on the part of parents, and not something that a coercive state forces on its citizens. In the words of one commentator, "The old eugenics would have required a continual selection for breeding of the fit, and a culling of the unfit. The new eugenics would permit in principle the conversion of all the unfit to the highest genetic level."[11]

Parents already make these kinds of choices when they discover through amniocentesis that their child has a high probability of Down's syndrome and decide to have an abortion. In the immediate future, the new eugenics is likely to lead to more abortions and discarded embryos, which is why those opposed to abortion will resist the technology strongly. But it will not involve coercion against adults, or restrictions on their reproductive rights. On the contrary, their range of reproductive choices will dramatically expand, as they cease to worry about infertility, birth defects, and a host of other problems. It is, moreover, possible to anticipate a time when reproductive technology will be so safe and effective that no embryos need be discarded or harmed.

My own preference is to drop the use of the loaded term *eugenics* when referring to future genetic engineering and substitute the word *breeding*—in German, *Züchtung*, the word originally used to translate Darwin's term *selection*. In the future, we will likely be able to breed human beings much as we breed animals, only far more scientifically and effectively, by selecting which genes we pass on to our children. *Breeding* has no necessary connotations of state sponsorship, but it is appropriately suggestive of genetic engineering's dehumanizing potential.

Any case to be made against human genetic engineering should therefore not get hung up on the red herring of state sponsorship or the prospect of government coercion. The old-fashioned eugenics remains a problem in authoritarian countries like China and may constitute a foreign policy problem for Western countries dealing with China.[12] But opponents of breeding new humans will have to explain what harms will be produced by the free decisions of individual parents over the genetic makeup of their children.

There are basically three categories of possible objections: (1) those based on religion; (2) those based on utilitarian considerations; and (3) those based on, for lack of a better term, philosophical principles. The remainder of this chapter will consider the first two categories of reservations, while Part II will deal with the philosophical issues.

## RELIGIOUS CONSIDERATIONS

Religion provides the clearest grounds for objecting to the genetic engineering of human beings, so it is not surprising that much of the opposition to a variety of new reproductive technologies has come from people with religious convictions.

In a tradition shared by Jews, Christians, and Muslims, man is created in God's image. For Christians in particular, this has important implications for human dignity. There is a sharp distinction between human and nonhuman creation; only human beings have a capacity for moral choice, free will, and faith, a capacity that gives them a higher moral status than the rest of animal creation. God acts through nature to produce these outcomes, and hence a violation of natural norms like having children through sex and the family is also a violation of God's will. While historical Christian institutions have not always acted on this principle, Christian doctrine emphatically asserts that all human beings possess an equal dignity, regardless of their outward social status, and are therefore entitled to an equality of respect.

Given these premises, it's not surprising that the Catholic Church and conservative Protestant groups have taken strong stands against a whole range of biomedical technologies, including birth control, in vitro fertilization, abortion, stem cell research, cloning, and prospective forms of genetic engineering. These reproductive technologies, even if freely embraced by parents out of love for their children, are wrong from this perspective because they put human beings in the place of God in creating human life (or destroying it, in the case of abortion). They allow reproduction to take place outside the context of the natural processes of sex and the family. Genetic engineering, moreover, sees a human being not as a miraculous act of divine creation, but rather as the sum of a series of material causes that can be understood and manipulated by human beings. All of this fails to respect human dignity, and thus violates God's will.

Given the fact that conservative Christian groups constitute the most visible and impassioned lobby opposed to many forms of reproductive technology, it is often assumed that religion constitutes the *only* basis on which one can be opposed to biotechnology and that the central issue is the question of abortion. While some scientists, like Francis Collins, the distinguished molecular biologist who since 1993 has headed the Human Genome Project, are observant Christians, the majority are not, and among this latter group there is a widespread view that religious conviction is tantamount to a kind of irrational prejudice that stands in the way of scientific progress. Some think that religious belief and scientific inquiry are incompatible, while others hope that greater education and scientific literacy will eventually lead to a withering away of religiously based opposition to biomedical research.

These latter views are problematic for a number of reasons. In the first place, there are many grounds to be skeptical about both the practical and ethical benefits of biotechnology that have nothing to do with religion, as Part II of this book will seek to demonstrate. Religion provides only the most straightforward motive for opposing certain new technologies.

Second, religion often intuits moral truths that are shared by non-religious people, who fail to understand that their own secular views on ethical issues are as much a matter of faith as those of religious believers. Many hardheaded natural scientists, for example, have a rational materialist understanding of the world, and yet in their political and ethical views are firmly committed to a version of liberal equality that is not all that different from the Christian view of the universal dignity of humankind. As will be seen below, it is not clear that the equality of respect for all human beings demanded by liberal egalitarianism flows logically from a scientific understanding of the world as opposed to being an article of faith.

Third, the view that religion will necessarily give ground to scientific rationalism with the progress of education and modernization more generally is itself extraordinarily naive and detached from empirical reality. It was the case that many social scientists a couple of generations ago believed that modernization necessarily implied secularization. But this pattern has been followed only in Western Europe; North America and Asia have seen no inevitable decline in religiosity with higher levels of education or scientific awareness. In some cases, belief in traditional religion has been replaced by belief in secular ideologies like "scientific" socialism that are no more rational than religion; in others, there has been a strong revival of traditional religion itself. The ability of modern societies to "free" themselves of authoritative accounts of who they are and where they are going is much more difficult than many scientists assume. Nor is it clear that these societies would necessarily be better off without such accounts. Given the fact that people with strong religious views are not likely to disappear from the political scene anytime soon in modern democracies, it behooves nonreligious people to accept the dictates of democratic pluralism and show greater tolerance for religious views.

On the other hand, many religious conservatives damage their own cause by allowing the abortion issue to trump all other considerations in biomedical research. Restrictions on federal funding for embryonic stem cell research were put in place by abortion opponents in Congress in 1995 to prevent harm to embryos. But embryos are routinely harmed by in vitro fertilization clinics when they are discarded, a practice that abortion opponents have been willing to let stand up to now. The National Institutes for Health had developed guidelines for conducting research in this extremely promising area without risk of raising the number of abortions performed in the United States. The guidelines mandated that embryonic stem cells should be derived not from aborted fetuses or those created specifically for research purposes, but from extra embryos produced as a by-product of in vitro fertilization, ones that would have been discarded or stored indefinitely were they not used in this fashion.[13] President George W. Bush modified these guidelines in 2001 by limiting federal funding to only those sixty or so stem cell "lines" (that is, cells that had been isolated and that could replicate indefinitely) that had already been produced. As Charles Krauthammer has pointed out, religious conservatives have focused on the wrong issue with regard to stem cells. They should not be worried about the sources of these cells but about their ultimate destiny: "What really ought to give us pause about research that harnesses the fantastic powers of primitive cells to develop into entire organs and even organisms is what monsters we will soon be capable of creating."[14]

While religion provides the most clear-cut grounds for opposing certain types of biotechnology, religious arguments will not be persuasive to many who do not accept religion's starting premises. We thus need to examine other, more secular, types of arguments.

## UTILITARIAN CONCERNS

By *utilitarian*, I mean primarily economic considerations—that is, that future advances in biotechnology may lead to unanticipated costs or long-term negative consequences that may outweigh the presumed benefits. The "harms" inflicted by biotechnology from a religious perspective are often intangible (for example, the threat to human dignity implied by genetic manipulation). By contrast, utilitarian harms are generally more broadly recognized, having to do either with economic costs or with clearly identifiable costs to physical well-being.

Modern economics provides us with a straightforward framework for analyzing whether a new technology will be good or bad from a utilitarian viewpoint. We assume that all individuals in a market economy pursue their individual interests in a rational fashion, based on sets of individual preferences that economists do not presume to judge. Individuals are free to do this as long as the pursuit of these preferences does not prevent other individuals from pursuing theirs; government exists to reconcile these individual interests through a series of evenhanded procedures embodied in law. We can further presume that parents will not seek to deliberately harm their children, but rather will try to maximize their happiness. In the words of the libertarian writer Virginia Postrel, "People want genetic technology to develop because they expect to use it *for themselves*, to help themselves and their children, to work and to keep their own humanity . . . In a dynamic, decentralized system of individual choice and responsibility, people do not have to trust any authority but their own."[15]

Assuming that the use of new biotechnologies, including technologies like genetic engineering, comes about as a matter of individual choice on the part of parents rather than being coercively mandated by the state, is it possible that harms can nonetheless result for the individual or for society as a whole?

The most obvious class of harms are the ones quite familiar to us from the world of conventional medicine: side effects or other long-term negative consequences to the individual undergoing treatment. The reason the Food and Drug Administration and other regulatory bodies exist is to prevent these kinds of harms, through the extensive testing of drugs and medical procedures before they are released on the market.

There is some reason to think that future genetic therapies, and particularly those affecting the germ line, will pose regulatory challenges significantly more difficult than those that have been experienced heretofore with conventional pharmaceuticals. The reason is that once we move beyond relatively simple single-gene disorders to behavior affected by multiple genes, gene interaction becomes very complex and difficult to predict (see Chapter 5, pp. 74–75). Recall the mouse whose intelligence was genetically boosted by neurobiologist Joe Tsien but which seems also to have felt greater pain as a result.

Given that many genes express themselves at different stages in life, it will take years before the full consequences of a particular genetic manipulation become clear.

According to economic theory, social harms can come about in the aggregate only if individual choices lead to what are termed negative externalities—that is, costs that are borne by third parties who don't take part in the transaction. For example, a company may benefit itself by dumping toxic waste in a local river but will harm other members of the community. A case like this has been made about Bt corn: it produces a toxin that kills the European corn borer, a pest, but it may also kill monarch butterflies. (This charge, it would appear, is not true.[16]) The issue is, Are there circumstances in which individual choices regarding biotechnology may entail negative externalities and thus lead to society as a whole being worse off?[17]

Children who are the subjects of genetic modification, obviously without consent, are the most clear class of potentially injured third parties. Contemporary family law assumes a community of interest between parents and children and therefore gives parents considerable leeway in the raising and educating of their offspring. Libertarians argue that since the vast majority of parents would want only what is best for their children, there is a kind of implied consent on the part of the children who are the beneficiaries of greater intelligence, good looks, or other desirable genetic characteristics. It is possible, however, to think of any number of instances in which certain reproductive choices would appear advantageous to parents but would inflict harm on their children.

### Politically Correct

Many kinds of characteristics that a parent might want to give a child have to do with the subtler elements of personality whose benefits are not as clear-cut as looks or intelligence. Parents may be under the sway of a contemporary fad or cultural bias or simple political correctness: one generation may prefer ultrathin girls, or pliable boys, or children with red hair—preferences that can easily fall out of favor in the next generation. One could argue that parents are already free to make such mistakes on behalf of their children and do so all the time by miseducating them or imposing their own quirky values on them.

But a child who is brought up in a certain way by a parent can rebel later. Genetic modification is more like giving your child a tattoo that she can never subsequently remove and will have to hand down not just to her own children but to all subsequent descendants.*

As noted in Chapter 3, we are already using psychotropic drugs to androgynize our children, giving Prozac to depressed girls and Ritalin to hyperactive boys. The next generation may for whatever reason prefer supermasculine boys and hyperfeminine girls. But you can always stop giving drugs to children if you don't like their effects. Genetic engineering, on the other hand, will embed one generation's social preferences in the next.

Parents can easily make wrong decisions concerning the best interests of their children because they rely on advice from scientists and doctors with their own agendas. The impulse to master human nature out of simple ambition or on the basis of ideological assumptions about the way people ought to be is all too common.

In his book *As Nature Made Him*, the journalist John Colapinto describes the heartbreaking story of a boy named David Reimer, who had the double misfortune of having his penis accidentally cauterized as a baby during a botched circumcision and falling under the supervision of a noted sex specialist at Johns Hopkins University, John Money. The latter stood at one extreme of the nature-nurture controversy, arguing throughout his career that gender identities are not natural but constructed after birth. David Reimer provided Money with an opportunity to test his theory, since he happened to be one of a pair of monozygotic twins and thus could be compared with his genetically identical twin brother. After the circumcision accident, Money had the boy castrated and oversaw the raising of David as a girl named Brenda.

Brenda's life became a private hell because she knew that, despite what her parents and Money told her, she was a boy and not a girl.

*It has been suggested that we will be able to sidestep the problem of consent in genetic engineering through the use of artificial chromosomes, which can be added to a child's normal genetic inheritance but switched on only after the child is old enough to be able to give his or her consent. See Gregory Stock and John Campbell, eds., *Engineering the Human Germline* (New York: Oxford University Press, 2000), p. 11.

From an early age she insisted on urinating standing up rather than sitting down. Later,

> Enrolled in Girl Scouts, Brenda was miserable. "I remember making daisy chains and thinking, If this is the most exciting thing in Girl Scouts, forget it," David says. "I kept thinking of the fun stuff my brother was doing in Cubs." Given dolls at Christmas and birthdays, Brenda simply refused to play with them. "What can you do with a doll?" David says today, his voice charged with remembered frustration. "You look at it. You dress it. You undress it. Comb its hair. It's boring! With a car, you can drive it somewhere, go places. I wanted cars."[18]

The effort to create a new gender identity wreaked so much emotional havoc that by the time Brenda reached puberty, she broke free of Money and had her sex change reversed through penis reconstruction; today David Reimer is reportedly a happily married man.

Nowadays it is much better understood that sexual differentiation begins well before birth, and that the brains of human males (as well as other animals) undergo a process of "masculinization" in utero when they receive a bath of prenatal testosterone. What is noteworthy about this story, however, is that Money could assert for almost fifteen years in scientific papers that he had succeeded in changing Brenda's sexual identity to that of a girl, when exactly the opposite was the case. Money was widely celebrated for his research. His fraudulent results were hailed by feminist Kate Millet in her book *Sexual Politics*, by *Time* magazine, and by *The New York Times* and were incorporated into numerous textbooks, including one in which they were cited as proving that "children can easily be raised as a member of the opposite sex" and that what few inborn sex differences might exist in humans "are not clear-cut and can be overridden by cultural learning."[19]

David Reimer's case stands as a useful warning about the uses to which biotechnology may be put in the future. His parents were driven by love for their child and desperation at the misfortune he had suffered, and they assented to a horrific treatment for which they felt profoundly guilty in later years. John Money was driven by a combination of scientific vanity, ambition, and the desire to make an ideological point, characteristics that led him to overlook contrary evidence and work directly against the interests of his patient.

Cultural norms may also lead parents to make choices that harm their children. One example was alluded to earlier, the use in Asia of sonograms and abortion to select the sex of offspring. In many Asian cultures, having a son confers clear-cut advantages to the parents in terms of social prestige and security for old age. But it clearly harms the girls who then fail to be born. Lopsided sex ratios also harm males as a group by making it harder for them to find appropriate mates and decreasing their bargaining position vis-à-vis females in marriage markets. If unattached males produce higher levels of violence and crime, then the society as a whole will suffer.

If we move from reproductive technologies to other aspects of biomedicine, there are additional types of negative externalities that can arise from rational individual decisions. One concerns aging and future prospects for life extension. Faced with a choice between dying and prolonging their lives through therapeutic intervention, most individuals will choose the latter, even if their enjoyment of life will be impaired to varying degrees as a result of the treatment. If large numbers of people make the choice to, for example, extend their lives for another ten years at the cost of, say, a 30 percent decrease in functionality, then society as a whole will have to pick up the tab for keeping them alive. This is, in effect, what has already begun to happen in countries that, like Japan, Italy, and Germany, have rapidly aging populations. One can imagine much more dire scenarios in which dependency ratios become even more extreme, leading to substantial declines in average standards of living.

The discussion of life extension in Chapter 4 suggests negative externalities that go beyond simple economic ones. The failure of older people to get out of the way will harm younger people seeking to move up the ladder in age-graded hierarchies. While any individual will want to postpone death as long as possible, people in the aggregate may not enjoy living in a society whose median age is 80 or 90, where sex and reproduction become activities engaged in by a small minority of the population, or where the natural cycle of birth, growth, maturity, and death has been interrupted. In one extreme scenario, the indefinite postponement of death will force societies to put severe constraints on the number of births allowed. Care for elderly parents has already begun to displace child care as a major preoccupation for people alive today. In the future, they may feel enslaved to the two, three, or more generations of ancestors dependent on them.

Another important type of negative externality is related to the competitive, zero-sum nature of many human activities and characteristics. Height confers many advantages on individuals who are above average, in terms of sexual attractiveness, social status, athletic opportunities, and the like. But these advantages are only relative: if many parents seek to have children tall enough to play in the NBA, it will lead to an arms race and no net advantage to those who participate in it.

This will even be true of a characteristic like intelligence, which is often cited as one of the first and most obvious targets of future genetic enhancement. A society with higher average intelligence may be wealthier, insofar as productivity correlates with intelligence. But the gains many parents seek for their children may prove illusory in other respects, because the advantages of higher intelligence are relative and not absolute.[20] People want smarter kids so that they will get into Harvard, for example, but competition for places at Harvard is zero-sum: if my kid becomes smarter because of gene therapy and gets in, he or she simply displaces your kid. My decision to have a designer baby imposes a cost on you (or rather, your child), and in the aggregate it is not clear that anyone is better off. This kind of genetic arms race will impose special burdens on people who for religious or other reasons do not want their children genetically altered; if everyone around them is doing it, it will be much harder to abstain, for fear of holding their own children back.

### Deference to Nature

There are good prudential reasons to defer to the natural order of things and not to think that human beings can easily improve on it through casual intervention. This has proven true with regard to the environment: ecosystems are interconnected wholes whose complexity we frequently don't understand; building a dam or introducing a plant monoculture into an area disrupts unseen relationships and destroys the system's balance in totally unanticipated ways.

So too with human nature. There are many aspects of human nature that we think we understand all too well or would want to change if we had the opportunity. But doing nature one better isn't always that easy; evolution may be blind process, but it follows a ruthless adaptive logic that makes organisms fit for their environments.

It is today politically correct, for example, to deplore human proclivities for violence and aggression, and to denounce the bloodlust that in earlier periods led to conquest, dueling, and similar activities. But there are some good evolutionary reasons such propensities exist. Understanding the good and bad in human nature is far more complex than one would think, because they are so intertwined. In evolutionary history, human beings learned, in biologist Richard Alexander's phrase, to cooperate in order to compete.[21] That is, the vast panoply of human cognitive and emotional characteristics that enable such an elaborate degree of social organization was created not by the struggle against the natural environment but rather by the fact that human groups had to struggle against one another. This led over evolutionary time to an arms-race situation, in which increasing social cooperation on the part of one group forced other groups to cooperate in similar ways in a never-ending struggle. Human competitiveness and cooperativeness remain balanced in a symbiotic relationship not just over evolutionary time, but in actual human societies and in individuals. We certainly hope that human beings will learn to live peacefully in many circumstances where they don't do so today, but if the balance shifts too far away from aggressive and violent behavior, the selective pressures in favor of cooperation will also weaken. Societies that face no competition or aggression stagnate and fail to innovate; individuals who are too trusting and cooperative make themselves vulnerable to others who are more bloody-minded.

So too with the family. Since Plato's time, it has been widely understood among philosophers that the family stands as the major obstacle to the achievement of social justice. People, as kin selection theory suggests, tend to love their families and relatives out of proportion to their objective worth. When there is a conflict between fulfilling an obligation to a family member and fulfilling an obligation to an impersonal public authority, family comes first. This is why Socrates argues in Book V of *The Republic* that a perfectly just city requires the communism of women and children, so that parents will not know who their biological offspring are and therefore will not be in a position to favor them.[22] This is also why all modern rule-of-law societies must enforce myriad regulations forbidding nepotism and favoritism in public service.

And yet the natural propensity to love one's own offspring to the point of irrationality has a powerful adaptive logic: if a mother does not love her children in this way, who else will devote the resources, both material and emotional, that are required to raise a child into mature adulthood? Other institutional arrangements, like communes and welfare agencies, work a good deal less well because they are not based on natural emotions. There is, moreover, a profound justice to the natural process, for it guarantees that even children who are unlovely or untalented will have a parent to love them in spite of their disadvantages.

Some have argued that even if we had the technological capability to change human personality in fundamental ways, we would never *want* to do so because human nature in some sense guarantees its own continuity. This argument, I believe, greatly underestimates human ambition and fails to appreciate the radical ways in which people in the past have sought to overcome their own natures. Precisely because of the irrationality of family life, all real-world communist regimes targeted the family as a potential enemy of the state. The Soviet Union celebrated a little monster named Pavel Morozov, who turned in his parents to Stalin's police in the 1930s, precisely to try to break the hold the family naturally has on people's loyalties. Maoist China engaged in a prolonged struggle against Confucianism, with its emphasis on filial piety, and turned children against parents during the Cultural Revolution in the 1960s.

It is impossible at this juncture to say how decisive any of these utilitarian arguments against certain developments in biotechnology will be. Much will depend on precisely how these technologies play out: whether we have life extension, for example, that does not simultaneously maintain a high quality of life, or develop genetic therapies that unexpectedly produce horrific effects that emerge only twenty years after first being administered.

The important point is this: we should be skeptical of libertarian arguments that say that as long as eugenic choices are being made by individuals rather than by states, we needn't worry about possible bad consequences. Free markets work well much of the time, but there are also market failures that require government intervention to correct. Negative externalities do not simply take care of themselves. We do not know at this point whether these externalities will be large or small, but we should not assume them away out of a rigid commitment to markets and individual choice.

### The Limitations of Utilitarianism

While it is convenient to argue for or against something on utilitarian grounds, all utilitarian arguments ultimately have a major limitation that often proves a decisive flaw. The goods and bads that utilitarians tote up in their cost-benefit ledgers are all relatively tangible and straightforward, usually reducible to money or to some easily identifiable physical harm to the body. Utilitarians seldom take into account more subtle benefits and harms that cannot be easily measured, or which accrue to the soul rather than to the body. It is easy to make a case against a drug like nicotine, which has clearly identifiable long-term health consequences, such as cancer or emphysema; it is harder to argue against a Prozac or a Ritalin, which may affect one's personality or character.

A utilitarian framework has particular difficulty encompassing moral imperatives, which tend to be regarded as just another type of preference. The University of Chicago economist Gary Becker, for example, argues that crime is the result of a rational utilitarian calculation: when the benefits of committing a crime outweigh the costs, a person will do so.[23] While this calculus is obviously what motivates many criminals, it implies at one extreme that people would be willing to, say, kill their own children if the price was right and they were assured of getting away with the crime. The fact that the vast majority of people would not ever think of entertaining such a proposition suggests that they in effect put an infinite value on their children, or that the obligation they feel to do right by them is not really commensurable with other types of economic values. There are, in other words, things that people believe to be morally wrong regardless of the utilitarian benefits that might flow from them.

So it is with biotechnology. While it is legitimate to worry about unintended consequences and unforeseen costs, the deepest fear that people express about technology is not a utilitarian one at all. It is rather a fear that, in the end, biotechnology will cause us in some way to lose our humanity—that is, some essential quality that has always underpinned our sense of who we are and where we are going, despite all of the evident changes that have taken place in the human condition through the course of history. Worse yet, we might make this change without recognizing that we had lost something of great value. We might thus emerge on the other side of a great divide between human and posthuman history and not even see that the watershed had been breached because we lost sight of what that essence was.

And what is that human essence that we might be in danger of losing? For a religious person, it might have to do with the divine gift or spark that all human beings are born with. From a secular perspective, it would have to do with human nature: the species-typical characteristics shared by all human beings qua human beings. That is ultimately what is at stake in the biotech revolution.

There is an intimate connection between human nature and human notions of rights, justice, and morality. This was the view held by, among others, the signers of the Declaration of Independence. They believed in the existence of natural rights, rights, that is, that were conferred on us by our human natures.

The connection between human rights and human nature is not clear-cut, however, and has been vigorously denied by many modern philosophers who assert that human nature does not exist, and that even if it did, rules of right and wrong have nothing whatever to do with it. Since the signing of the Declaration of Independence, the term *natural rights* has fallen out of favor and has been replaced with the more generic *human rights*, whose provenance does not depend on a theory of nature.

It is my view that this turn away from notions of rights based on human nature is profoundly mistaken, both on philosophical grounds and as a matter of everyday moral reasoning. Human nature is what gives us a moral sense, provides us with the social skills to live in society, and serves as a ground for more sophisticated philosophical discussions of rights, justice, and morality. What is ultimately at stake with biotechnology is not just some utilitarian cost-benefit calculus concerning future medical technologies, but the very grounding of the human moral sense, which has been a constant ever since there were human beings. It may be the case that, as Nietzsche predicted, we are fated to move beyond this moral sense. But if so, we need to accept the consequences of the abandonment of natural standards for right and wrong forthrightly and recognize, as Nietzsche did, that this may lead us into territory that many of us don't want to visit.

To survey this terra incognita, however, we need to understand modern theories of rights and what role human nature plays in our political order.

**PART II**

·

BEING HUMAN

7

# HUMAN RIGHTS

**Terms like sanctity remind me of animal rights. Who gave a dog a right? This word right gets very dangerous. We have women's rights, children's rights; it goes on forever. And then there's the right of a salamander and a frog's rights. It's carried to the absurd.**

**I'd like to give up saying rights or sanctity. Instead, say that humans have needs, and we should try, as a social species, to respond to human needs— like food or education or health—and that's the way we should work. To try and give it more meaning than it deserves in some quasi-mystical way is for Steven Spielberg or somebody like that. It's just plain aura, up in the sky—I mean, it's crap.**

**James Watson[1]**

I f James Watson, Nobel laureate, discoverer of the structure of DNA, and one of the towering figures of twentieth-century science, should become a bit impatient with the injection of the word *rights* into the discourse of his particular domain of genetics and molecular biology, we might well excuse him. Watson is famous both for his temper and for his often unguarded and politically incorrect remarks; he is, after all, a hardheaded scientist and not a scribbler on political and social matters. Moreover, he is correct in his scatological observation about contemporary rights discourse. His remark is reminiscent of the words of the utilitarian philosopher Jeremy Bentham, who famously commented that the French Declaration of the Rights of Man and the Citizen's assertion that rights were natural and imprescriptible was "nonsense upon stilts."

The problem, however, does not end there, because we cannot in the end dispense with serious discussion of rights and talk only of needs and interests. Rights are the basis of our liberal democratic political order and key to contemporary thinking about moral and ethical issues. And any serious discussion of human rights must ultimately be based on some understanding of human ends or purposes, which in turn must almost always rest on a concept of human nature. And it is here that Watson's field, biology, becomes relevant, because the life sciences have been making important discoveries about human nature in recent years. As much as natural scientists would like to maintain a Chinese wall of separation between the natural "is" that they study and the moral and political "ought" engendered in discourse on rights, this is ultimately a dodge. The more science tells us about human nature, the more implications there are for human rights, and hence for the design of institutions and public policies that protect them. These findings suggest, among other things, that contemporary capitalist liberal democratic institutions have been successful because they are grounded in assumptions about human nature that are far more realistic than those of their competitors.

## RIGHTS TALK

Over the past generation, the rights industry has grown faster than an Internet IPO in the late 1990s. In addition to the aforementioned animal, women's, and children's rights, there are gay rights, the rights of the disabled and handicapped, indigenous people's rights, the right to life, the right to die, the rights of the accused, and victims' rights, as well as the famous right to periodic vacations that is laid out in the Universal Declaration of Human Rights. The U.S. Bill of Rights is reasonably clear in enumerating a certain set of basic rights to be enjoyed by all American citizens, but in 1971 the Supreme Court, in *Roe v. Wade*, manufactured a new right out of whole cloth, based on Justice Douglas's finding of a right to abortion that was an "emanation" from the "penumbra" of the similarly shadowy right to privacy in the earlier *Griswold v. Connecticut* decision. The constitutional scholar Ronald Dworkin, in his book *Life's Dominion*, comes up with an even more novel argument: since having an abortion is a major life decision on a par with making a religious commitment, the right to abortion turns out all along to have been protected by the First Amendment's guarantee of religious liberty.[2]

The situation becomes even more confused when discussion of rights turns to futuristic issues like genetic technology. Bioethicist John Robertson, for example, argues that individuals have a fundamental right to what he calls procreative liberty, which involves both a right to reproduce as well as a right not to reproduce (including, therefore, a right to abortion). But the right to reproduce is not limited to reproduction through coital means (that is, through sex); it also applies to reproduction through noncoital means like in vitro fertilization. Quality control, it turns out, is protected by the same right, and hence "genetic screening and selective abortion, as well as the right to select a mate or a source for donated eggs, sperm, or embryos, should be protected as part of procreative liberty."[3] It may come as a surprise to many that they have a fundamental right to do something that is not, as yet, fully possible technologically, but such is the wonderfully elastic nature of contemporary rights talk.

Ronald Dworkin, for his part, proposes what amounts to a right to genetically engineer people, not so much on the part of parents but of scientists. He posits two principles of "ethical individualism" that are basic to a liberal society—the first, that each individual life be successful rather than wasted, and second, that while each life is equally important, the person whose life it is has special responsibility for its outcome. On this basis, he argues that "if playing God means struggling to improve what God deliberately or nature blindly has evolved over eons, then the first principle of ethical individualism commands that struggle, and its second principle forbids, in the absence of positive evidence of danger, hobbling scientists and doctors who volunteer to lead it."[4]

Given all of this monumental confusion over what constitutes a right and where they come from, why don't we follow James Watson's advice and abandon talk of rights altogether, and simply speak of human "needs" or "interests"? Americans more than most peoples have tended to conflate rights and interests. By transforming every individual desire into a right unconstrained by community interests, one increases the inflexibility of political discourse. The debates in the United States over pornography and gun control would appear much less Manichean if we spoke of the interests of pornographers rather than their fundamental First Amendment right to free speech, or the needs of owners of assault weapons rather than their sacred Second Amendment right to bear arms.

## THE NECESSITY OF RIGHTS

So why not abandon what legal theorist Mary Ann Glendon labels rights talk altogether? The reason we cannot do so, as either a theoretical or practical matter, is that the language of rights has become, in the modern world, the only shared and widely intelligible vocabulary we have for talking about ultimate human goods or ends, and in particular, those collective goods or ends that are the stuff of politics. Classical political philosophers like Plato and Aristotle did not use the language of rights—they spoke of the human good and human happiness, and the virtues and duties that were required to achieve them. The modern use of the term *rights* is more impoverished, because it does not encompass the range of higher human ends envisioned by the classical philosophers. But it is also more democratic, universal, and easily grasped. The great struggles over rights since the American and French revolutions are testimony to the political salience of this concept. The word *right* implies moral judgment (as in, "What is the right thing to do?") and is our principal gateway into a discussion of the nature of justice and of those ends we regard as essential to our humanity.

Watson is in effect advocating a utilitarian approach with his advice to simply try to satisfy human needs and interests without reference to rights. But this runs into the typical problem of utilitarianism: the question of priorities and justice when needs and interests conflict. A powerful and important community leader is in need of a new liver because of a drinking problem; I am an indigent, terminally ill patient in a public hospital, on life support but with a healthy liver. Any simple utilitarian calculation that seeks to maximize the satisfaction of human needs would dictate that I be involuntarily removed from life support so that my liver can be harvested for the sake of the important leader and the people who depend on him. That no liberal society permits this to happen reflects a view that innocent people have a *right* not to be involuntarily deprived of life, no matter how many important needs may be satisfied as a result.

Let us consider another example, much less pleasant to contemplate, to illustrate the limits of utilitarianism. One of the less appetizing aspects of the contemporary food chain, usually hidden from the view of the consumers of food, is the process of rendering. All of the beef, chickens, pigs, lambs, and other types of animals that we eat are, of course, slaughtered and turned into hamburgers, roasts, chicken sandwiches, and the like. Once the parts edible by humans are processed, however, there remains a huge volume of animal carcasses, amounting to millions of tons of organic matter each year, that needs to be disposed of. Hence the modern rendering industry, which takes these carcasses and chops, shreds, or boils them down into further usable products, such as oil, bonemeal, and finally food products that are fed back to animals. We force cows and other animals, in other words, to be cannibals.*

Why, on utilitarian grounds, do we not similarly render the dead bodies of human beings and turn them into animal feed or some other useful product, assuming this can be done with the deceased's consent? Why cannot people be permitted to donate their bodies voluntarily not just to scientific research but to be reprocessed into food? One could argue on utilitarian grounds that the economic value of the body of a typical elderly dead person is not very high, but surely there are more cost-effective ways of disposing of it than uselessly warehousing it in the ground in perpetuity. There must be many poor families that would benefit greatly from the few dollars that could be

---

*It is believed that bovine spongiform encephalopathy (BSE, or mad cow disease) was transmitted in this manner: the proteinlike prions that cause this disease in the brains of infected animals were not destroyed in the rendering process, but were preserved in animal feed and fed to healthy animals.

obtained by selling off the body parts of a dead brother or father killed in an urban gun battle. Reasoning along similar lines, what sense does it make for soldiers to risk their lives to recover the body of a fallen comrade? Why do families waste precious resources trying to recover the body of a missing child or brother?

The reason that we do not begin to contemplate alternatives such as the rendering of human bodies—the reason that just the articulation of such a possibility arouses immediate feelings of disgust—has to do with the words that James Watson disliked using, such as *sanctity* and *dignity*. That is, we attach an extraordinary noneconomic value to the bodies of the dead and feel that they need to be treated with a respect not due the carcass of a cow because they are *human* bodies. A utilitarian might argue in response that these feelings of disgust or respect are simply parts of the pains and pleasures on which utilitarian calculations are made. But this simply begs the further question of why human beings, in a species-typical way, invest each other with such special emotions, emotions that extend even to the lifeless bodies of relatives and loved ones.

Rights trump interests because they are endowed with greater moral significance. Interests are fungible and can be traded off against one another in a marketplace; rights, while seldom absolute, are less flexible because it is hard to assign them an economic value. I may have an interest in a pleasant two-week vacation, but that cannot compete with another person's right not to be held as a chattel slave to work someone else's fields. The slave's right to freedom is not just a strong interest on the slave's part; a disinterested third party might say that the condition of servitude is unjust because it is an affront to the slave's dignity as a human being. The slave's freedom is somehow more basic and fundamental to his status as a human being than my interest in a pleasant vacation is to mine, even if I might assert my interest more passionately than the slave does his.

Political systems enshrine certain kinds of rights over others, and thereby reflect the moral basis of their underlying societies. The United States was founded on the principle stated in the Declaration of Independence, that "all men are created equal, that they are endowed by their Creator with certain unalienable rights." That principle, as Abraham Lincoln explained, was violated by the institution of slavery and necessitated the fighting of a bloody civil war. This then paved the way for the Emancipation Proclamation and passage of the Fourteenth Amendment, which corrected this grave inconsistency and laid the basis for subsequent American democracy.

So if rights prioritize human ends or goods and set some above others as the foundation of justice, where do they come from? The reason there is a constant inflation in the scope of rights is precisely that everyone wants to raise the relative priority of certain interests over others. In the cacophony of rights talk, how do we decide what is genuinely a right and what is not?

Rights derive in principle from three possible sources: divine rights, natural rights, and what one might call contemporary positivistic rights, located in law and social custom. Rights, in other words, can emanate from God, Nature, and Man himself.

Rights derived from revealed religion are not today the acknowledged basis of political rights in any liberal democracy. John Locke begins his *Second Treatise on Government* with an attack on Robert Filmer and the doctrine of divine right; the very essence of modern liberalism was to eliminate religion as the explicit basis of political order. This was based on a practical observation that religiously based polities were constantly at war with one another because there was not sufficient consensus on religious first principles. The background for Hobbes's description of the state of nature as a war of "every man against every man" was the sectarian violence of his time. None of this, of course, prevents private individuals in liberal societies from believing that man is a creature created in God's image and that basic human rights therefore come from God. Such views become problematic only when they are asserted as political rights, as in the abortion controversy. For they then run into the same problem recognized by Locke: it is extremely difficult to achieve political consensus on issues involving religion.

The second possible source of rights is nature, or more precisely, human nature. Despite Jefferson's invocation of the Creator in the Declaration, he believed, like Locke and Hobbes, that rights needed to be grounded in a theory of human nature. A political principle like equality had to be based on empirical observation of what human beings were like "by nature." The practice of slavery was in principle contrary to nature and therefore unjust.

The idea that human rights can be based on human nature has been vigorously attacked from the eighteenth century to the present. This attack has gone under the label of the naturalistic fallacy, a tradition that stretches from David Hume to twentieth-century analytical philosophers such as G. E. Moore, R. M. Hare, and others.[5] Particularly strong in the Anglo-Saxon world, the naturalistic fallacy argues that nature cannot provide a philosophically justifiable basis for rights, morality, or ethics.[6]

Since the philosophical school dominant in contemporary academia believes that any attempt to base rights on nature has long since been debunked, it is understandable that natural scientists are quick to invoke the naturalistic fallacy as a shield to protect their work from unpalatable political implications of the sort laid out in Chapter 2. Since most natural scientists are either apolitical or else *bien-pensant* liberals, it is easy for them to evoke the naturalistic fallacy and argue, as Paul Ehrlich recently did in his book *Human Natures*,[7] that human nature gives us absolutely no guidance as to what human values should be.

It is my view that the common understanding of the naturalistic fallacy is itself fallacious and that there is a desperate need for philosophy to return to the pre-Kantian tradition that grounds rights and morality in nature. But before I can state this argument more fully and explain why the dismissal of natural rights is misguided, we need to look at the third source of rights, which is what might be labeled as positivistic. The weaknesses of the third, positivistic approach to rights are indeed what necessitate the effort to resurrect the concept of natural rights.

The simplest way to locate the source of rights is to look around and see what society itself declares to be a right, through its basic laws and declarations. William F. Schultz, executive director of Amnesty International, argues that contemporary human rights advocates have long since dropped any notion that human rights can or should be based on nature or natural law.[8] Instead, according to him, " 'human rights' refers to 'humans' rights,' 'the rights of humans,' something that humans can possess or can claim, but not necessarily something derived from the nature of the claimant." Human rights are, in other words, whatever human beings say they are.

If you take his statement as a political strategy for negotiating documents like the Universal Declaration of Human Rights, there is no doubt that Schultz is correct in saying that rights are whatever you can get people to agree they are, and that there will never be consensus on a set of natural rights. There can be procedural refinements to make sure a positive right actually reflects the will of the society that declares it, such as rules that require ratification of bills of rights by supermajorities (as in the case of the U.S. Constitution). The First Amendment rights to freedom of speech and religion may or may not be ordained by nature, but they are ratified as part of a constitutional process. But this approach means that rights are essentially procedural: if you can get a supermajority (or whatever) to agree that all people have a right to walk around in public in their underwear, this then becomes a fundamental human right along with freedom of association and freedom of speech.

So what's wrong with a purely positivistic approach to rights? The problem, as any human rights advocate will know in practice, if not in theory, is that there are no positive rights that are also universal. When Western human rights groups criticize the Chinese government for jailing political dissidents, the Chinese government responds that for its society, collective and social rights outweigh individual rights. The emphasis of Western organizations on individual political rights is not an expression of a universal aspiration, but rather reflects the Western (or perhaps Christian) cultural biases of the human rights groups themselves. The Western human rights advocate might respond that the Chinese government hasn't followed the correct procedure, insofar as it hasn't consulted its own people in a democratic manner. But if there are no universal standards for political behavior, who is to say what the right procedure is? And what does an advocate of a positivistic approach like rights campaigner William Schultz have to say in response to another, culturally different society, that follows the proper procedures yet promotes an abhorrent practice like suttee or slavery or female circumcision? The answer is that no response is possible, since it has been declared from the outset that there are no transcendent standards for determining right and wrong beyond whatever the culture declares to be a right.

## WHY THE NATURALISTIC FALLACY
## IS FALLACIOUS

The problem of cultural relativism brings us back to reconsider whether we might have been premature in discarding an approach to human rights based on human nature, since the existence of a single human nature shared by all the peoples of the world can provide, at least in theory, a common ground on which to base universal human rights. Belief in the naturalistic fallacy runs so deep in contemporary Western thought, however, that resurrection of a natural rights argument remains a formidable task.

The idea that rights cannot be grounded in nature rests on two separate though often interrelated arguments. The first is attributed to David Hume, one of the fathers of British empiricism, who is widely believed to have proven once and for all that it is impossible to derive an "ought" from an "is." In a famous passage from his *Treatise of Human Nature*, Hume notes,

> In every system of morality with which I have hitherto met, I have always remark'd, that the author proceeds for some time in the ordinary way of reasoning, and establishes the being of a God, or makes observations concerning human affairs; when of a sudden I am surpriz'd to find, that instead of the usual copulations of propositions, *is*, and *is not*, I meet with no proposition that is not connected with an *ought*, or an *ought not*. This change is imperceptible; but is, however, of the last consequence. For as this *ought* or *ought not*, expresses some new relation or affirmation, 'tis necessary that it should be observ'd and explain'd; and at the same time that a reason should be given, for what seems altogether inconceivable, how this new relation can be a deduction from others, which are entirely different from it.[9]

Hume is usually credited with asserting that a statement of moral obligation cannot be derived from an empirical observation about nature or the natural world. When natural scientists assert that their work has no political or policy implications, they usually have in mind the Humean is-ought dichotomy: because human beings are genetically inclined to behave in certain species-typical ways does not imply that they *should* behave in that manner. Moral obligation comes from some other shadowy, ill-defined realm distinct from the natural world.

The second strand of the naturalistic fallacy would argue that even if we could derive an "ought" from an "is," the "is" is often ugly, amoral, or indeed immoral. Anthropologist Robin Fox argues that biologists have learned a great deal about human nature in recent years, but that it is not very pleasant to behold, and would serve poorly as a basis for political rights.[10] Evolutionary biology, for example, has given us the theory of kin selection, or inclusive fitness, which asserts that human beings seek to maximize their reproductive fitness by favoring genetic relatives in proportion to their shared genes. This leads, in Fox's view, to the following implications:

> A very good argument could be made, using basic kin selection theory, that there is a natural and human right to revenge. If someone kills my nephew or grandson, he robs me of a proportion of my inclusive fitness, that is, the strength of my personal gene pool. To redress this imbalance, it could be argued, I have the right to inflict similar loss on him . . . This system of vengeance is less efficient than a redress system whereby I would get to impregnate one of the perpetrator's females, thus forcing him to raise to viability a person carrying my own genes.[11]

To rebuild an argument in favor of natural right, we need to take on each of these arguments in turn, beginning with the is-ought distinction. More than forty years ago, the philosopher Alasdair MacIntyre pointed out that Hume himself neither believed in nor abided by the rule commonly attributed to him that one could not derive an "ought" from an "is."[12] At most, what the famous passage from the *Treatise* said was that one could not deduce moral rules from empirical fact in a logically a priori way. *But like virtually every serious philosopher in the Western tradition* since Plato and Aristotle,[13] Hume believed that the "ought" and the "is" were bridged by concepts like "wanting, needing, desiring, pleasure, happiness, health"—by the goals and ends that human beings set for themselves. MacIntyre gives the following example of how one is derived from the other: "If I stick a knife in Smith, they will send me to jail; but I do not want to go to jail; so I ought not (had better not) stick a knife in him."

There are, of course, a huge variety of human wants, needs, and desires that can produce an equal diversity of "oughts." Why do we not end up back at utilitarianism, which in effect creates moral "oughts" by seeking to satisfy human needs? The problem with utilitarianism in its various forms lies not with its method of bridging the "is" and the "ought": many utilitarians base their ethical principles on explicit theories of human nature. The problem rather lies in utilitarianism's radical reductionism—that is, in the overly simplified view of human nature that utilitarians employ.[14] Jeremy Bentham sought to reduce all human motivation to the pursuit of pleasure and the escape from pain;[15] more modern utilitarians such as B. F. Skinner and the behavioralists had a similar concept in mind when they talked of positive and negative reinforcement. Modern neoclassical economics begins from a model of human nature that posits that human beings are rational utility maximizers. Economists explicitly disavow any attempt to distinguish between or prioritize individual utilities; in fact, they often reduce all human activities, from those of a Wall Street investment banker to Mother Teresa ministering to the poor, to a pursuit of indistinguishable units of consumer preference called utiles.*

There is an elegant simplicity to the reductionist strategy underlying utilitarian ethics, which explains its appeal to many. It promises that ethics can be transformed to something like a science, with clear-cut rules for optimization. The problem is that human nature is far too complex to be reduced to simple categories like "pain" and "pleasure." Some pains and pleasures are deeper, stronger, and more abiding than others. The pleasure we derive from reading a trashy pulp-fiction novel is different from the pleasure of reading *War and Peace* or *Madame Bovary* with the benefit of life experiences of the sort that these latter novels address. Some pleasures point us in contradictory directions: a drug addict may crave rehabilitation and a drug-free life at the same time that he wants his next fix.

We could see more clearly the way human beings actually bridge

---

*In Mother Teresa's case, the utile would have to be some form of psychological satisfaction.

the "is" and the "ought" by recognizing that human values are intimately bound up, as a matter of empirical fact, with human emotions and feelings. The "oughts" thereby derived are at least as complex as the human emotional system. That is, there is scarcely a judgment of "good" or "bad" that has been pronounced by a human being that has not been accompanied by a strong emotion, whether of desire, longing, aversion, disgust, anger, guilt, or joy. Some of these emotions encompass the simple pains and pleasures of the utilitarians, but others reflect more complex social feelings, such as the desire for status or recognition, pride in one's ability or righteousness, or shame at having violated a social rule or prohibition. When we unearth the tortured body of a political prisoner in an authoritarian dictatorship, we pronounce the words *bad* and *monstrous* because we are driven by a complex gamut of emotions: horror at the decomposed body, sympathy for the victim's sufferings and those of family and friends, and anger at the injustice of the killing. We may temper these judgments by rational consideration of mitigating circumstances: perhaps the victim was a member of an armed terrorist group; perhaps counterinsurgency requires the government to take repressive measures that claim innocent victims. But the process of value derivation is not fundamentally a rational one, because its sources are the "is" of the emotions.

All emotions are by definition experienced subjectively; how then do we move to a more objective theory of value when they come into conflict with one another? It is at this point that traditional philosophical accounts of human nature enter the picture. Virtually every pre-Kantian philosopher had an implicit or explicit theory of human nature that set certain wants, needs, emotions, and feelings above others as more fundamental to our humanness. I may want my two-week vacation, but your desire to escape slavery is based on a more universal and more deeply felt longing for freedom, and it therefore trumps my want. Hobbes's assertion of a basic right to life (which is the precursor to the right to life enshrined in the Declaration of Independence) is based on an explicit theory of human nature that posits that the fear of violent death is the strongest of human passions and therefore produces a right more basic than, say, the assertion of religious orthodoxy. The moral opprobrium that attaches to murder is due in large measure to the fact that the fear of death is part of human nature and does not vary substantially from one human community to another.

One of the earliest philosophical accounts of human nature is that given by Socrates in Plato's *Republic*. Socrates argues that there are three parts to the soul: a desiring part (*eros*), a spirited or prideful part (*thymos*), and a rational part (*nous*). These three parts are not reducible to one another and are in many ways not commensurable: my *eros* or desire might tell me to break ranks and run back from the battlefield to my family, but my *thymos* or pride leads me to stand fast for fear of shame. Different conceptions of justice favor different parts of the soul (democracy, for example, favors the desiring part, while aristocracy favors the spirited part), and the best city satisfies all three. Because of this three-part complexity, even the most just city requires that some parts of the soul cannot be fully satisfied (like the famous communism of women and children that effectively abolishes the family), and no real-world political system can hope to achieve more than an approximation of justice. Yet justice remains a meaningful concept, whose plausibility stands or falls on the plausibility of the underlying three-part psychology from which it is derived. (Many thoughtless contemporary commentators sneer at Plato's "simplistic" psychology that divides the soul in three, without realizing that many twentieth-century schools of thought, including Freudianism, behavioralism, and utilitarianism, are even more simpleminded, reducing the soul only to its desiring element, with reason playing no more than an instrumental role, and *thymos* out of the picture entirely.)

The radical break in the Western tradition comes not with Hume but with Rousseau, and particularly with Kant.[16] Rousseau, like Hobbes and Locke, sought to characterize man in the state of nature, but he also argued in the *Second Discourse* that human beings were "perfectible"—that is, that they had the capability to alter their natures over time. Perfectibility provided the seed for Kant's idea of a noumenal realm that was free of natural causation and that was the ground of the categorical imperative, which detached morality in its entirety from any concept of nature. Kant argued that we had to assume the existence of the possibility of true moral choice and freedom of the will. By definition, moral action could not be the product of a natural desire or instinct but had to act *against* natural desire on the basis of what reason alone dictates to be right. According to his famous statement at the beginning of the *Foundations of the Metaphysics of Morals*, "Nothing in the world—indeed, nothing even beyond the world—can possibly be conceived which could be called good without qualification except a *good will*."[17] All other characteristics or ends desired by human beings, from intelligence and courage to wealth and power, were good only relative to the goodness of the will that possessed them; a good will was the only thing desirable in itself. Kant posited that qua moral agents, human beings were noumena, or things-in-themselves, that therefore had always to be treated as ends rather than as means.

A number of observers have pointed out the similarities between Kantian ethics and the view of human nature embodied in Protestantism, which holds that the latter is irredeemably sinful and that moral behavior requires rising above or suppressing our natural desires in toto.[18] Aristotle and the medieval Thomistic tradition argued that virtue built upon and extended what nature provided us, and that there was no necessary conflict between what was naturally pleasurable and what was right. In Kantian ethics, we see the beginnings of the view that the good is a matter of the will *overcoming* nature.

Much of subsequent Western philosophy has followed the Kantian route toward so-called deontological theories of right—that is, theories that try to derive a system of ethics that is not dependent on any substantive assertions about human nature or human ends. Kant himself said that his moral rules would apply to any rational agents, even if they were not human beings; society could in fact be composed of "rational devils." Following Kant, subsequent deontological theories begin from the premise that there can be no substantive theory about human ends, whether drawn from human nature or any other source.

According to John Rawls, for example, in a liberal state, "systems of ends are not ranked in value";[19] individual "life plans" can be distinguished by their greater or lesser rationality, but not by the nature of the goals or ends they set.[20] This is the view that has become embedded in a good deal of thinking on contemporary U.S. constitutional law. Post-Rawlsian legal theorists like Ronald Dworkin and

Bruce Ackerman try to define the rules of a liberal society while es-
chewing any reference to priorities among human ends or, in more
contemporary language, between possible lifestyles.[21] Dworkin has argued that the liberal state "must be neutral on . . . the question of the
good life . . . political decisions must be, so far as is possible, independent of any particular conception of the good life, or of what gives
value to life." Ackerman, for his part, asserts that no social arrangement can be justified "if it requires the power holder to assert (a) that
his conception of the good is better than that asserted by any of his
fellows, or (b) that, regardless of his conception of the good, he is in-
trinsically superior to one or more of his fellow citizens."[22]

I believe that this broad turn away from human nature–based theories of right is flawed for a number of reasons. Perhaps the most revealing weakness of deontological theories of right is that virtually all
philosophers who attempt to lay out such a scheme end up reinsert-
ing various assumptions about human nature into their theories. The
only difference is that they do it covertly and dishonestly, rather than
explicitly, as in the earlier tradition from Plato to Hume. William Galston points out how Kant himself, in *The Metaphysical Elements of
Justice*, asserts that a community cannot impose on itself an ecclesias-
tical constitution in which certain religious dogmas are held to be
permanent, because such an arrangement "would conflict with the
appointed aim and purpose of mankind." And what is the purpose of
mankind? To develop as rational individuals, free of obscurantist prejudice. This assertion of Kant's already makes several strong assumptions about human nature: that humans are rational creatures, that
they benefit from and enjoy the use of their rationality, and that they
may develop this rationality over time. The latter implies the need for
education, and for a state that is not neutral on the question of
whether citizens can choose dogmatic ignorance or education.

The same is true of contemporary Kantians like John Rawls,
whose theory of justice explicitly sidesteps any discussion of human
nature and seeks to establish a set of minimal moral rules that would
apply to any group of rational agents, based on the so-called original
position. That is, we are to select rules of just distribution from behind a "veil of ignorance," where we don't know what our actual position in society is. As critics of Rawls have pointed out, the original

position itself, and the political implications Rawls draws from it, contains numerous assertions about human nature, in particular his assumption that human beings are risk-averse.[23] He assumes they would choose a strictly egalitarian distribution of resources for fear of ending up on the bottom of the social ladder. But many individuals may in fact prefer a more hierarchical society, taking the risk of ending up with low status for a chance at reaching high status. Moreover, Rawls spends a good deal of time in *A Theory of Justice* elaborating the conditions under which human beings can optimally establish plans, which at a minimum assumes that they are purposive, rational animals that can formulate long-term goals. And he often makes appeals to what are in effect observations about human nature, as in the following passage:

> The basic idea is one of reciprocity, a tendency to answer in kind. Now this tendency is a deep psychological fact. Without it our nature would be very different and fruitful social cooperation fragile if not impossible . . . Beings with a different psychology either have never existed or must soon have disappeared in the course of evolution.[24]

The assertion that reciprocity is both genetically programmed as part of human psychology and necessary for the survival of human beings as a species should have significant implications for the moral status of reciprocity as a form of ethical behavior.

Ronald Dworkin similarly asserts that "it is objectively important that any human life, once begun, succeed rather than fail—that the potential of that life be realized rather than wasted."[25] This single phrase abounds with assumptions about human nature: that each human life has a distinct natural potential; that that potential is something which develops over time; that whatever this potential is requires some effort and foresight to cultivate; and that there are preferences and choices that an individual can have or make relative to that potential that would be less than desirable, from both the individual's standpoint and that of the larger society. A truly deontological theory would assert that if a large number of individuals in a society spent the first half of their lives earning money so that they could spend the second half in a heroin stupor, and violated no procedural rules in the process, that would be fine: there is no substantive theory of human nature or of the good that would allow us to distinguish between a person who actively sought to improve himself or herself through education and participation in society, and a drug addict. Obviously, neither Rawls nor Dworkin believes this, which means that they cannot escape making certain judgments about what is naturally best for human beings.

There is no better illustration of the way that covert or backdoor human nature theorizing reasserts itself than in the writing of bioethicist John Robertson, who, as noted earlier, has postulated a right of "procreative liberty," which in turn is said to entail a right to the genetic modification of one's offspring. Where does the right of procreative liberty come from, since it is nowhere to be found in the Bill of Rights? Surprisingly, Robertson does not base it on positive law, such as the rights to privacy and abortion that were established by the Supreme Court in the *Griswold v. Connecticut* and *Roe v. Wade* decisions. Rather, he simply invents the right on the following grounds:

> Procreative liberty should enjoy presumptive primacy when conflicts about its exercise arise because control over whether one reproduces or not is central to personal identity, to dignity, and to the meaning of one's life. For example, deprivation of the ability to avoid reproduction determines one's self-definition in the most basic sense. It affects women's bodies in a direct and substantial way. It also centrally affects one's psychological and social identity and one's social and moral responsibilities. The resulting burdens are especially onerous for women, but they affect men in significant ways as well.
>
> On the other hand, being deprived of the ability to reproduce prevents one from an experience that is central to individual identity and meaning in life. Although the desire to reproduce is in part socially constructed, at the most basic level transmission of one's genes through reproduction is an animal or species urge closely linked to the sex drive. In connecting us with nature and future generations, reproduction gives solace in the face of death.[26]

Phrases like "central to personal identity" and "self-definition in the most basic sense," as well as references to the body being affected "in a direct and substantial way," all suggest priorities among the wide variety of human desires and purposes. They make the case that purposes connected with reproduction constitute basic rights because they are somehow more important than other kinds of aims, based on their importance for a median or average human individual. Not all people feel strongly about reproductive decisions—for certainly there are people who don't want to reproduce or for whom the decision to have a child isn't a big deal. But the *typical* human being does care about such things. Indeed, Robertson overtly appeals to nature, saying that "transmission of one's genes through reproduction is an animal or species urge." One is tempted to paraphrase Hume: one is surpriz'd to remark an almost imperceptible shift on the part of deontological writers from *ought* and *ought not* to *is* and *is not*, since they no more than anyone else can avoid basing what "ought" to be on what typically "is" for our species.

Modern deontological theories of right have other weaknesses. In default of a substantive theory of human nature or any other means of grounding human ends, deontological theories wind up elevating individual moral autonomy to the highest human good. They offer individuals the following bargain: neither philosophers nor society in the form of the liberal state will tell you how to live your life, but rather they let *you* decide. All that either will do is to establish some procedural rules to ensure that your chosen life plan doesn't interfere with the life plans of your fellow citizens. This explains the great popularity of this approach: no one likes having his life plan criticized or denigrated. The right to choose, rather than inherently meaningful life plans, is the only thing that deontological theories consistently protect. As the plurality opinion in the 1992 Supreme Court decision *Casey v. Planned Parenthood* put it, "at the heart of liberty is the right to define one's own concept of existence, of meaning, of the universe, and of the mystery of human life."[27]

Much in contemporary culture supports the view that moral autonomy is the most important human right. The germ of this idea comes from Kant's view that human beings are noumena or things-in-themselves capable of moral freedom. From Nietzsche comes the view that man is the "beast with red cheeks"—a value creator who is able to will values into existence by pronouncing the words *good* and *bad* and applying these words to the world around him. From here it

is a short step to the values discourse of contemporary democratic societies, where I am totally free to make up my own values regardless
of whether they are shared more broadly by others in the larger community.[28]

But while freedom to choose one's own plan for life is certainly a
good thing, there is ample reason to question whether moral freedom
as it is currently understood is such a good thing for most people, let
alone the single most important human good. The kind of moral autonomy that has traditionally been said to give us dignity is the freedom to accept or reject moral rules that come from sources higher
than ourselves, and not the freedom to make up those rules in the
first place. For Kant, moral autonomy didn't mean following your
personal inclination wherever it led, but rather obedience to the a
priori rules of practical reason, which forced us often to do things at
cross-purposes with our natural individual desires and inclinations.
Contemporary understandings of individual autonomy, by contrast, seldom provide a way to distinguish between genuine moral choices and
choices that amount to the pursuit of individual inclinations, preferences, desires, and gratifications.

Even if we accept at face value the claim that individual choice
constitutes moral autonomy, the primacy of the ability to make limit-
less choices over other human goods is not self-evident. Some people
may favor life plans that defy authority and tradition and break commonly accepted social rules. But other life plans can be fulfilled only
in conjunction with other people, and these require limitations of individual autonomy for the sake of social cooperation or community
solidarity. A perfectly plausible life plan may entail living in a traditional religious community (say, of Mennonites or Orthodox Jews),
which then seeks to restrict the personal freedom of the community's
members. Another life plan may involve living in a tightly bonded
ethnic community, or living a life of republican virtue in which all
individualism gives way to life in the barracks. Ethics based on deontological principles is not truly neutral among life plans; it favors
the more individualistic ones that predominate in liberal societies
over more communitarian ones that may be just as humanly satisfying.

Human beings have been wired by evolution to be social creatures

who naturally seek to embed themselves in a host of communal relationships.* Values are not arbitrary constructs but serve an important
purpose in making collective action possible. Human beings also find
great satisfaction in the fact that values and norms are *shared*. Solip-
sistically held values defeat their own purpose and lead to a highly
dysfunctional society in which people are unable to work together for
common ends.

What about the other leg of the naturalistic fallacy argument,
which says that even if rights were derivable from nature, that nature
is violent, aggressive, cruel, or indifferent? Human nature at a minimum points in contradictory directions, toward competition and
cooperation, toward individualism and sociability; how can any particular "natural" behavior be the basis of natural rights?

The answer, I believe, is that while there is no *simple* translation of
human nature into human rights, the passage from one to the other is
ultimately mediated by the rational discussion of human ends—that
is, by philosophy. That discussion does not lead to a priori or mathe-
matically provable truths; indeed, it may not even yield substantial
consensus among the discussants. It does, however, allow us to begin
to establish a hierarchy of rights and, importantly, allows us to rule
out certain solutions to the problem of rights that have been politically powerful in the course of human history.

Take for example the human propensity for violence and aggression. Few would deny that this is somehow grounded in human nature;
there are virtually no societies free of murder or that have not experienced armed violence in some form. But what we notice in the first
instance is that random violence against other members of the community is prohibited in every known human cultural group: while murder is universal, so are laws and/or social norms that seek to prohibit
murder. This is no less true among man's primate cousins: a troop of
chimpanzees will occasionally experience violent aggression from a
younger male that, like the Columbine High School shooters, is lonely,
peripheral, or seeking to make a point.[29] But the older members of the
community will always take measures to control and neutralize that individual because community order cannot tolerate such violence.

*This point will be defended more fully in the following chapter.

Primate violence, including human violence, is legitimated primarily at higher social levels—that is, on the part of in-groups that compete with out-groups. Warriors are treated with respect and honor in a way that school shooters are not. Hobbes's war of "every man against every man" is in fact a war of every group against every group. In-group social order is driven by the need for competition against out-groups, both over evolutionary time (there is a great deal of evidence that human cognitive capabilities were shaped by these group-oriented competitive needs[30]) and in the course of human history.[31] There is a sad continuity from nonhuman primates to hunter-gatherer societies to contemporary participants in ethnic and sectarian violence as (primarily) male-bonded groups compete against one another for dominance.[32]

This might be taken to be confirmation of the naturalistic fallacy and thus the end of the story, except for the fact that human nature encompasses a great deal more than male-bonded violence. It also involves the desire for what Adam Smith called gain, the accumulation of property and goods useful to life, as well as reason, the capacity for foresight and the rational ordering of priorities over the long term. When two human groups butt up against each other, they face a choice between engaging in a violent, zero-sum struggle for dominance, or else in a peaceful, positive-sum relationship of trade and exchange. Over time, the logic of the latter choice (what Robert Wright labels nonzero-sumness[33]) has driven the boundaries of human in-groups to ever-larger communities of trust: from tiny kin groups to tribes or lineages, to states, nations, broad ethnolinguistic communities, and what Samuel Huntington labels cultures—communities of shared values encompassing many nation-states and hundreds of millions, if not billions, of people.

There remains a significant amount of violence at the boundaries of these ever-larger groups, made more deadly by the simultaneous advance of military technology. But there is a logic to human history that is ultimately driven by the priorities that exist among natural human desires, propensities, and behaviors. Human violence has over the past 100,000 years become increasingly controlled and pushed to the outer boundaries of these ever-larger groups. Globalization—a world order in which mankind's largest in-groups no longer violently compete with one another for dominance but trade peaceably—can be seen as the logical culmination of a long-term series of decisions in favor of positive-sum competition.

Violence, in other words, may be natural to human beings, but so is the propensity to control and channel violence. These conflicting natural tendencies do not have equal status or priority; human beings reasoning about their situation can come to understand the need to create rules and institutions that constrain violence in favor of other natural ends, such as the desire for property and gain, that are more fundamental.

Human nature also serves to provide us with guidance as to what political orders won't work. Proper understanding of the contemporary evolutionary theory of kin selection, or inclusive fitness, for example, would have led us to predict the bankruptcy and ultimate failure of communism, due to the latter's failure to respect the natural inclination to favor kin and private property.

Karl Marx argued that man is a species-being: that is, that human beings have altruistic feelings toward the human species as a whole. The policies and institutions of real-world communist states, like the abolition of private property, the subordination of the family to the party-state, and the commitment to universal worker solidarity, were all predicated on this belief.

There was a time when evolutionary theorists like V. C. Wynne-Edwards postulated the existence of species-level altruism, but modern kin selection theory argues against the existence of strong group-selection pressures.[34] It postulates instead that altruism arises primarily out of the need of *individuals* to get their genes passed on to successive generations. Human beings will by this account be altruistic primarily to family members and other kin; a political system that forces them to spend their Saturdays away from their families, working on behalf of the "heroic Vietnamese people," will meet with very deep resistance.

The preceding example demonstrates the ways in which human nature and politics are intertwined: kin selection indicates that a political system that respects the right of people to follow their own individual self-interests and attend to family and close friends before they attend to strangers halfway around the world will be more stable, workable, and satisfying than one that does not. Human nature does not dictate a single, precise list of rights; it is both complex and flexible as it interacts with various natural and technological environments. But it is not infinitely malleable, and our underlying shared humanity allows us to rule out certain forms of political order, like tyranny, as unjust. Human rights that speak to the most deeply felt and universal human drives, ambitions, and behaviors will be a more solid foundation for political order than those that do not. This explains why there are a lot of capitalist liberal democracies around the world at the beginning of the twenty-first century but very few socialist dictatorships.

It is thus impossible to talk about human rights—and therefore justice, politics, and morality more generally—without having some concept of what human beings actually are like as a species. To assert this is not to deny that History in the Hegelian-Marxist sense exists.[35] Human beings are free to shape their own behavior because they are cultural animals capable of self-modification. History has brought about huge changes in human perceptions and behavior such that a member of a hunter-gatherer society and the inhabitant of a contemporary information society seem in many respects to belong to different species. Evolving human institutions and cultural arrangements have produced different human moral attitudes over time. But nature puts limits on the kinds of self-modification that have hitherto been possible. In the words of the Latin poet Horace, "You can throw Nature out with a pitchfork/But it always comes running back." There will still be a glimmer of recognition when the tribesman and the Internet maven meet.

So if human rights rest on a substantive concept of nature, what is that concept? Can it be defined in a way that does justice to everything that is known scientifically about human behavior? Up to this point, I have not put forward a theory of human nature, or even a definition of what human nature is. There are many—most commonly in the social sciences, but among natural scientists as well—who would deny that human nature exists in any meaningful way. Hence we need, in the chapter that follows, to examine what a species-typical behavior is, and what it might be for our species.

8

# HUMAN NATURE

'According to nature' you want to *live*? O you noble Stoics, what deceptive words these are! Imagine a being like nature, wasteful beyond measure, indifferent beyond measure, without purposes and consideration, without mercy and justice, fertile and desolate and uncertain at the same time; imagine indifference itself as a power—how *could* you live according to this indifference?

Friedrich Nietzsche, *Beyond Good and Evil*, Section 9

I have up to this point presented the argument that human rights are properly based on human nature without defining what I mean by the term. Due to the intimate connection that exists between human nature, values, and politics, it is perhaps not surprising that the very concept of human nature has been extraordinarily controversial over the past couple of centuries. Most traditional discussions have revolved around the age-old question of where to draw the boundary line between nature and nurture. This argument was replaced by a different polemic in the late twentieth century, in which the balance shifted strongly in favor of nurture arguments, with many arguing strongly that human behavior was so plastic as to make human nature a meaningless concept. While recent progress in the life sciences has made the latter position less and less tenable, the anti–human nature position continues to live on: the environmentalist Paul Ehrlich recently expressed the hope that people would abandon all talk of human nature once and for all because it was a meaningless concept.[1]

The definition of the term *human nature* I will use here is the following: human nature is the sum of the behavior and characteristics that are typical of the human species, arising from genetic rather than environmental factors.

The word *typical* requires some explanation. I use the term in the same way that ethologists do when they speak of "species-typical behavior" (for example, pair-bonding is typical of robins and catbirds but not of gorillas and orangutans). One common misunderstanding about the "nature" of an animal is that the word implies rigid genetic determination. In fact, all natural characteristics show considerable variance within the same species; natural selection and evolutionary adaptation could not occur were this not so. This is particularly the case with cultural animals like human beings: since behaviors can be learned and modified, the variance in behavior is inevitably greater and will reflect the individual's environment to a greater extent than for animals incapable of cultural learning. This means that typicality is a statistical artifact—it refers to something close to the median of a distribution of behavior or characteristics.

Take human height. There is, obviously, considerable variance in human heights; within any given population, heights will exhibit what statisticians label a normal (bell curve–shaped) distribution. If we were to plot male and female heights for the United States today, they would look something like Figure 1 (the lines are meant to be illustrative only).

These curves tell us a number of things. There is, first of all, no such thing as a "normal" height; the distribution of heights in a population does, however, have a median and a mean.* Strictly speaking, there is no such thing as a "species-typical" height, only a species-typical distribution of heights; we all know that there are dwarfs and

---

*The median is the height at which half the population is taller and half shorter; the mean is the average height of the whole population.

**Height Distributions, 2000**

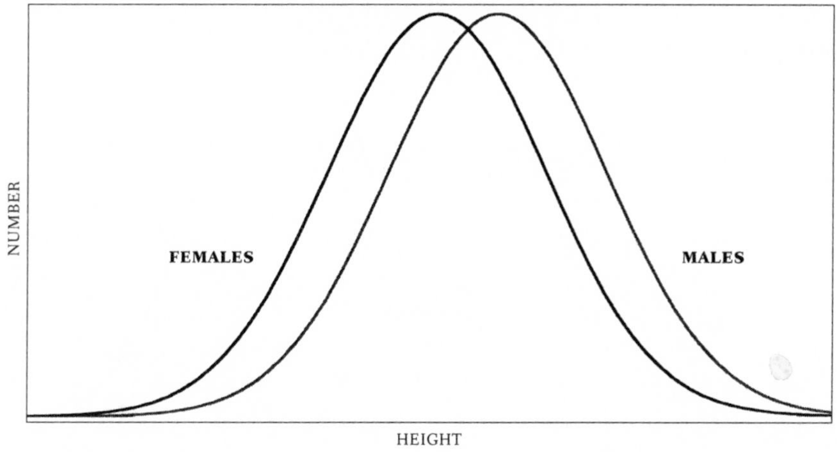

FIGURE 1

giants. There is also no strict definition of a dwarf or giant; a statistician might say arbitrarily that dwarfism begins two or more standard deviations below the mean, and giantism a similar number above. Neither dwarfs nor giants like being characterized as such, since these words carry a connotation of abnormality and stigma, and in ethical terms there is no reason to stigmatize them. But none of this means that it is meaningless to talk about species-typical heights for a population of human beings: the median of the human distribution will be different from the median of the distributions for chimpanzees and elephants, and the shape of the bell curve—the degree of variance—may differ as well. Genes play a role in determining both the medians and the shapes of the curves; they are also responsible for the fact that the medians of the male and female curves differ from each other.

But the way that nature and nurture interact is in fact a good deal more complicated. The median heights of different human groups vary considerably not just by sex, but by race and ethnic group. A lot of this is due to environment: the median height of Japanese in generations past was considerably lower than that of Europeans, but in the period after World War II, it increased with different and better

**Height Distributions over Time**

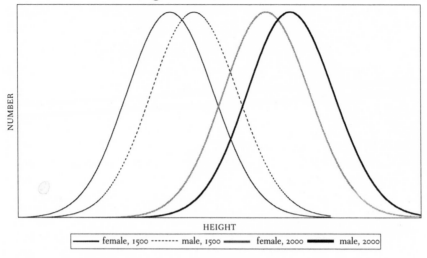

**FIGURE 2**

diets. In general, with economic development and improved nutrition, median heights have risen all around the world. If we compare the height distributions for a typical European country between the years 1500 and 2000, we come up with a set of curves something like the ones in Figure 2.

Nature, then, does not establish a single median human height; median heights are themselves normally distributed depending on diet, health, and other environmental factors. There has been a great increase in average heights since the Middle Ages, as is obvious to any museum visitor looking at the suits of armor worn by medieval knights. On the other hand, there are limits to the degree of variance possible, limits that are set genetically: if you deprive a population of enough calories on average, they starve to death rather than growing shorter, while past a certain point, increasing calorie intake makes them fatter, not taller. (This, needless to say, is the situation in much of the developed world today.) The average European woman in the year 2000 was considerably taller than the average man in the year 1500; but men remain on the whole taller than women. The actual medians for any given population or historical period are determined in large measure by the environment; but the overall degree of possible variance and the average male-female differences are the products of heredity and thus nature.

It may strike some that such a statistical definition of human nature is at variance with either the commonplace understanding of the term or the concept of human nature as employed by Aristotle and other philosophers. It is in fact only a more precise use of the term. When we observe someone taking a bribe and shake our heads with the remark, "It's human nature to betray people's trust," or when Aristotle asserts, as he does in the *Nicomachean Ethics*, that man is "a political animal by nature," the implication is never that *all* people take bribes or that *all* people are political. We all know of individuals who are honest or who are hermits; the assertion about human nature is either a probabilistic one (that is, an assertion about what most people will do most of the time) or else a conditional statement about how people are likely to interact with their environment ("If faced with easy temptations, most people will take bribes").

## CONTRA NATURAM

There are three broad categories of argument that critics have put forward over the years to make the case that the traditional concept of human nature is misleading or refers to something that does not exist. The first has to do with the claim that there are no true human universals that can be traced to a common nature, and that the ones that do exist are trivial (for example, the fact that all cultures prefer health to sickness).

The ethicist David Hull argues that many of the traits said to be universal to humans and uniquely characteristic of our species are in fact neither. This includes even language:

> Human language is not universally distributed among human beings. Some human beings neither speak nor understand anything that might be termed a language. In some sense such people might not be 'truly' human, but they still belong to the same biological species as the rest of us . . . They are potential language-users in the sense that if they had a different genetic make-up and were exposed to the ap-

propriate sequences of environments, then they would have been able to acquire language skills similar to those possessed by the rest of us. But this same contrary-to-fact condition can be applied to other species as well. In this same sense, chimpanzees possess the capacity to acquire language.[2]

Hull goes on to point out that there are any number of characteristics of a species that do not distribute themselves normally, and which therefore cannot be described in terms of a single median and standard deviation. Blood types are an example: an individual is an O, A, B, AB, and so on, but never an intermediate type between O and A. The types correspond to distinct alleles within human DNA, which either are or are not expressed, like switches that can be turned on or off. Certain blood types may be more or less prevalent within particular populations, but because they do not form a continuum (like differing heights), it is meaningless to speak of a species-typical blood type. Other characteristics distribute themselves on a continuum: skin color, for example, varies from light to dark but clusters by racial group around a series of peaks or modes.

This argument against the existence of human universals is specious because it uses too narrow a definition of *universal*. It is true that one cannot talk about a "universal" or median blood type, because blood types are what statisticians call categorical variables—that is, a characteristic that falls into a number of unordered, distinct categories. Nor does it make sense to talk about a "typical" skin color. But many other characteristics, such as height and strength, as well as psychological traits such as intelligence, aggressiveness, and self-esteem, fall along a continuum and distribute themselves normally around a single median point within any given population. The degree to which a population varies around this median (known as its standard deviation) is a measure, in some sense, of how typical the median is; the smaller the standard deviation, the more typical the median point.

This is the context in which a concept like "human universals" must be understood. A characteristic does not need to have a variance (standard deviation) of zero to be considered a universal, since almost none exist.[3] There are doubtless some mutant female kangaroos born without pouches, and some bulls born with three horns on their heads. Facts like these do not render meaningless the assertion that pouches are somehow constitutive of "kangarooness," or that bulls are creatures that typically have two horns on their heads.[4] For a characteristic to be considered universal, it needs rather to have a single, distinct median or modal point, and a relatively small standard deviation—something like curve I in Figure 3.

The second criticism of the concept of human nature is the one that has been put forward repeatedly over the years by geneticist Richard Lewontin,[5] to the effect that the genotype of an organism (its DNA) does not fully determine its phenotype (the actual creature that ultimately develops from the DNA). That is, even our physical appearance and features, not to mention our mental condition and behavior, are shaped by our environments rather than by heredity. Genes interact with environment at virtually every level of an organism's development and therefore determine much less than is usually asserted by proponents of the concept of human nature.

We have already seen an example of this in the case of median heights, which are partly determined by nature and partly by diet and other nutritional factors. Lewontin illustrates his point with a number

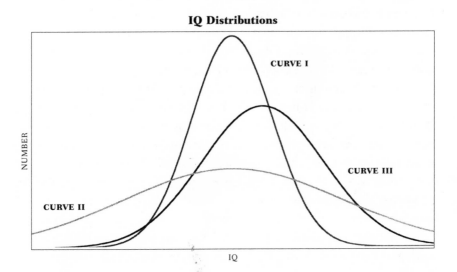

**IQ Distributions**

FIGURE 3

of other examples. He points out that even mice bred to be genetically identical will respond differently to poison in the environment, and that the fingerprints of identical twins are never identical.[6] There is a species of plant that grows in the mountains, whose outward appearance changes completely depending on the altitude at which it grows. It is well known that two babies with the same genetic endowment will turn out physically and mentally quite different from each other based on the behavior of the mother while each baby is in utero—whether the mother drinks, takes drugs, gets adequate nourishment, and so on. An individual's interaction with its environment therefore starts well before birth; characteristics we tend to attribute to nature are, by this argument, the product of a complex nature-environment interaction.

This particular reprise of the nature-nurture controversy can be illustrated by differently shaped distribution curves. For example, the tall curve I in Figure 3 is a hypothetical distribution of IQs in a population, under the (unrealistic) assumption that all individuals faced identical environments with regard to such factors affecting IQ as nutrition, education, and the like. This represents natural or genetic variance. The actual distribution of IQs in any population will inevitably be more like curve II, reflecting the fact that society harms some and benefits others in ways that affect intelligence. The curve is shorter and squatter, with more individuals at a greater distance from the median. The greater the difference in shape between the two curves, the more impact environment has relative to heredity.

Lewontin's argument is valid as far as it goes but hardly vitiates the concept of human nature. As noted in the discussion of height, environment can change median heights, but it cannot push human heights above or below certain limits, nor can it make women on average taller than men. Those parameters are still set by nature. Moreover, there is very often a linear relationship between environment, genotype, and phenotype that ensures that if the genetic variance is distributed normally, the phenotypic variance will also be distributed normally. That is, the better our diets, the taller we tend to be (within our species-typical limits); height distribution curves still have single median points despite the fact that they are affected by environment. Most human characteristics do *not* resemble the mountain plant that looks entirely different depending on elevation. Babies do not grow fur if raised in a cold climate, or gills if they live near the sea.

The important argument thus is not whether environment affects the kind of behavior and characteristics that are typical of the human species, but by how much. Chapter 2 noted Murray and Herrnstein's assertion in *The Bell Curve* that as much as 70 percent of the variance in IQs is due to heredity rather than environment. Lewontin and his colleagues have argued that the actual figure is significantly lower than this, to the point that hereditary factors, for them, ultimately play a very small role in determining IQ.[7] This is an empirical issue, and one where Lewontin appears to be wrong: the consensus of the discipline of psychology, relying on twin studies, maintains that though the figure is lower than the Murray-Herrnstein estimate, it is still in the range of 40 to 50 percent.

The degree to which a trait or behavior is heritable will vary greatly; preferences in music are almost entirely shaped by environment, which has almost no effect on a genetic disease like Huntington's chorea. Knowing the degree of heritability of a specific trait is very important if the trait is a significant one like IQ: those individuals in the area above curve I but below curve II were presumably put there not by nature but by their environment. If that area is large, there is greater hope of being able to move the curve's median to something more like that of curve III through some combination of diet, education, and social policy.

While Lewontin's arguments that genotypes do not determine phenotypes apply to all species, the third category of criticism of the concept of a species-typical nature is one that applies almost exclusively to human beings.[8] To wit, humans are cultural animals who can modify their own behavior based on learning, and pass that learning on to future generations in nongenetic ways.[9] This means that the variance in human behavior is far greater than for virtually any other species: human kinship systems range from elaborate clans and lineages down to single-parent families, in a way that the kinship systems of gorillas and robins do not. According to an anti–human nature polemicist like Paul Ehrlich, our nature is not to have a single nature. Thus he argues that "citizens of long-standing democracies have different human natures from those accustomed to living under dictatorships," while at another point he observes that "the natures of many Japanese people changed greatly in response to the defeat and the revelation of Japanese war crimes."[10] This reminds one of the memorable phrase in one of Virginia Woolf's novels that "on or about December, 1910, human character changed."

Ehrlich is simply restating an extreme form of the social-constructionist view of human behavior that was widely held fifty years ago but that has been progressively undermined by new research in recent decades. It is true that popular press coverage of "genes for" everything from breast cancer to aggression have given people a false sense of biological determinism, and it is useful to be reminded that culture and social construction continue to play important roles in our lives. But the finding that IQ is 40 to 50 percent heritable already contains within it an estimate of the impact of culture on IQ and implies that even taking culture into account, there is a significant component of IQ that is genetically determined.

The argument that human nature does not exist because human beings are cultural animals capable of learning is fundamentally misguided because it tilts against a straw man. No serious theorist of human nature ever denied that humans were cultural creatures, or that they could use learning, education, and institutions to shape the way they live. Aristotle maintained that human nature does not automatically lead us to our forms of flourishing in the way that an acorn grows into an oak tree. Human flourishing depends on the virtues, which human beings must deliberately acquire: "The virtues, therefore, are engendered in us neither by nature nor yet in violation of nature; nature gives us the capacity to receive them, and [this is] brought to perfection by habit."[11] This variability in individual development is mirrored by a variability in the rules of justice: despite the fact that there is such a thing as natural justice, "all rules of justice are variable."[12] The perfection of justice required that someone establish cities, and the writing of laws for those cities that suited existing conditions.[13] Aristotle notes that while "the right hand is naturally stronger than the left, yet it is possible for any man to make himself ambidextrous": culture supplements and can overcome nature. There is plenty of room in Aristotle's system, then, for what we today call cultural variation and historical evolution.

Both Plato and Aristotle held that reason was not simply a set of cognitive capabilities given to us at birth. Rather, it represented a kind of endless striving for knowledge and wisdom that needed to be cultivated in youth through education, and in later life through the accumulation of experience. Human reason did not dictate a single set of institutions or a best way to live in what Kant would later label an a priori fashion (that is, in the manner of a mathematical proof). It did, however, permit human beings to enter into a philosophical consideration of the nature of justice or of the best way to live based both on their unchanging natures *and* on their changing environment. The open-ended character of the human striving for knowledge was fully compatible with a concept of human nature—indeed, it constituted for the classical political philosophers a critical part of what they understood human nature to be.

## SO WHAT IS HUMAN NATURE, ANYWAY?

The life sciences have added a great deal to our store of empirical knowledge about human behavior and human nature, and it is a worthwhile enterprise to revisit some of the classical accounts of human nature. We can then see which ones stand up under the weight of new evidence, which seem to have been disproved, and which need to be modified in light of what we now know. A number of scholars have attempted to do this already, including Roger Masters,[14] Michael Ruse and Edward O. Wilson,[15] and Larry Arnhart.[16] Arnhart's book, *Darwinian Natural Right*, attempts to show that Darwin does not undermine Aristotle's ethical system and that the results of contemporary Darwinian biology can be used to support many of Aristotle's claims about natural morality.[17] Arnhart lists twenty natural desires that are universals characterizing human nature.[18]

Such lists are likely to be controversial; they tend either to be too short and general, or overly specific and lacking in universality. More important than a comprehensive definition for our present purposes is an effort to zero in on characteristics that are unique to the species, since these are critical to any understanding of the ultimate question of human dignity. We can begin with cognition, a species characteristic of which we humans tend to be inordinately proud.

### The Tabula Rasa Filled In

Much of what we have learned in recent years about human nature concerns, as we will see below, the species-typical ways in which we perceive, learn, and develop intellectually. Human beings have their own mode of cognition, which is different from that of apes and dolphins, one which is open-ended in the knowledge that it can accumulate, but is not infinitely so.

An obvious example of this is language. Actual human languages are conventional, and one of the greatest gulfs that separates one human group from another is the mutual unintelligibility of different languages. On the other hand, the ability to *learn* languages is universal and governed by certain biological characteristics of the human brain. In 1959, Noam Chomsky suggested that there were "deep structures" underlying the syntax of all languages;[19] the idea that these deep structures are innate, genetically programmed aspects of brain development is widely accepted today.[20] It is genes and not culture that ensure that the ability to learn languages appears at some point in the first year of child development and then diminishes by the time a child reaches adolescence.

The idea that there are innate forms of human cognition is one that has received a tremendous amount of empirical support in recent years but has also met a great deal of resistance. The reason for this resistance, particularly in the Anglo-Saxon world, is due to the lasting influence of John Locke and the school of British empiricism that he fostered. Locke begins *An Essay Concerning Human Understanding* with the assertion that there are no innate ideas in the human mind and, in particular, no innate moral ideas. This is the famous Lockean tabula rasa: the brain is a kind of general-purpose computer that can take in and manipulate the sensory data that appear to it. But its memory banks are essentially blank at the moment of birth.

Locke's tabula rasa remained a powerfully appealing idea through the middle of the twentieth century, when it was taken up by the behavioralist school of John Watson and B. F. Skinner. The latter advanced an even more radical version, to the effect that there were no species-specific modes of learning, and that pigeons, for example, could be made to recognize themselves in a mirror the way apes and humans could, given the proper rewards and punishments.[21] Modern cultural anthropology also accepts the tabula rasa assumption; anthropologists have argued, among other things, that the concepts of time and color are social constructions not present in every culture.[22] A great deal of the research emphasis in this field and in the related area of cultural studies over the past two generations has been to seek out the unusual, bizarre, or unexpected in human cultural practices, under the Lockean presumption that a single exception to a general rule will invalidate the rule.

Today the idea of the tabula rasa lies in shambles. Research in cognitive neuroscience and psychology has replaced the blank slate with a view of the brain as a modular organ full of highly adapted cognitive structures, most of them unique to the human species. There are in fact what amount to innate ideas or, more accurately, innate species-typical forms of cognition, and species-typical emotional responses to cognition.

The problem with Locke's view of innate ideas is partly definitional: he argues that nothing can be either innate or universal if it is not shared by every single individual in a population. Using the statistical language from the beginning of this chapter, he argues in effect that a natural or innate characteristic must have no variance or a standard deviation of zero. But as we have seen, nothing in nature exhibits this characteristic: even two monozygotic twins with identical genotypes will show *some* variance in their phenotypes due to slightly different conditions in utero.

The case Locke makes against the existence of moral universals suffers from a similar weakness in its demand for zero variance.* He argues that the Golden Rule (that is, the principle of reciprocity), which is a key precept of Christianity and other world religions, is not honored by all people, and is violated by many in practice.[23] He notes that even the love of parents for their children and vice versa does not

---

*Locke gets caught in another definitional problem, which is that he wants to talk about innate ideas in the strict sense of a verbal proposition, such as "Parents, preserve your children." He argues that the implicit statement about duty cannot be understood without a concept of law and lawmakers. It is true that there are no universal ideas in this form; what is universal are the human emotions that impel parents to protect their children and seek the best for them. The further step of articulating the values implied by these emotions does not always occur.

prevent enormities like infanticide and the deliberate killing of elderly parents.[24] Infanticide, he observes, has been practiced without remorse by the Mingrelians, Greeks, Romans, and other societies.

But while explicit linguistic formulations of the Golden Rule may not be universal in human cultures, there are no cultures that do not practice some type of reciprocity, and few that fail to make it a central component of moral behavior. A strong case can be made that this is not simply the result of learned behavior. The work of biologist Robert Trivers has shown that some form of reciprocity is evident not just across different human cultures but in the behavior of a range of nonhuman animal species as well, indicating that it has genetic causes.[25] Similarly, basic kin selection theory explains the evolutionary emergence of parental love.

There have been a number of ethological studies of infanticide in recent years, showing that it is practiced widely in the animal world as well as in a variety of human cultures.[26] None of these, however, prove Locke's point, because the more closely one looks at the actual practice of infanticide, the more it becomes clear that it is motivated by exceptional circumstances that explain how the normally powerful emotions of parental care can be overridden.[27] These circumstances include the desire of a stepfather or new mate to eliminate a rival's offspring; desperation, sickness, or extreme poverty on the part of the mother; a cultural preference for males; and an infant that is sickly or deformed. It is hard to find societies in which infanticide is not practiced primarily by those at the bottom of the social hierarchy; where resources permit families to raise their children, nurturing instincts dominate. And contrary to Locke, even when infanticide occurs, it is seldom practiced "without regret."[28] Infanticide is thus like murder more broadly considered: something that occurs universally but is universally condemned and controlled.

There is, in other words, a natural human moral sense that evolved over time out of the requirements of hominids, who were to become an intensely social species. Locke is right about the blank slate in the narrow sense that we are not born with preformed abstract moral ideas. There are, however, innate human emotional responses that guide the formation of moral ideas in a relatively uniform way across the species. These in turn are part of what Kant labeled the transcendental unity of apperception—that is, human ways of perceiving reality that give those perceptions order and meaning. Kant believed that space and time were the only inevitable structures of human apperception, but we can add a number of others to the list. We see colors, react to smells, recognize facial expressions, parse language for evidence of deceit, avoid certain dangers, engage in reciprocity, pursue revenge, feel embarrassment, care for our children and parents, feel repulsion for incest and cannibalism, attribute causality to events, and many other things as well, because evolution has programmed the human mind to behave in these species-typical ways. As in the case of language, we must learn to exercise these capabilities by interacting with our environment, but the potential for developing them, and the ways in which they are programmed to develop, are there at birth.

## HUMAN SPECIFICITY AND THE
## RIGHTS OF ANIMALS

The connection between rights and species-typical behavior becomes obvious when we consider the issue of animal rights. There is today around the world a very powerful animal rights movement, which seeks to improve the lot of the monkeys, chickens, minks, pigs, cows, and other animals that we butcher, experiment on, eat, wear, turn into upholstery, and otherwise treat as means rather than ends in themselves. The radical fringe of this movement has on occasion turned violent, bombing medical research labs and chicken processing plants. The bioethicist Peter Singer has built his career around the promotion of animal rights and a critique of what he calls the speciesism of human beings—the unjust favoring of our species over others.[29] All of this leads us to raise the question posed by James Watson at the beginning of Chapter 7: What gives a salamander a right?

The simplest and most straightforward answer to this question, which applies perhaps not to salamanders but certainly to creatures with more highly developed nervous systems, is that they can feel pain and suffer.[30] This is an ethical truth to which any pet owner can testify, and much of the moral impulse behind the animal rights movement is understandably driven by the desire to reduce the suffering of animals. Our greater sensitivity to this issue stems in part from the general spread of the principle of equality in the world, but also from an accumulation of greater empirical knowledge about animals.

Much of the work done in animal ethology over the past few generations has tended to erode the bright line that was once held to separate human beings from the rest of the animal world. Charles Darwin, of course, provided the theoretical underpinning for the notion that man evolved from an ancestral ape, and that all species were undergoing a continuous process of modification. Many of the attributes that were once held to be unique to human beings—including language, culture, reason, consciousness, and the like—are now seen as characteristic of a wide variety of nonhuman animals.[31]

For example, the primatologist Frans de Waal points out that culture—that is, the ability to transmit learned behaviors across generations through nongenetic means—is not an exclusively human achievement. He cites the famous example of the potato-washing macaques that inhabit a small island in Japan.[32] In the 1950s a group of Japanese primatologists observed that one macaque in particular (an Albert Einstein, so to speak, among monkeys) developed a habit of washing potatoes in a local stream. This same individual later discovered that grains of barley could be separated from sand by dropping them in water. Neither was a genetically programmed behavior; neither potatoes nor barley were part of the macaques' traditional diet, and no one had ever before observed these behaviors taking place. Yet both the potato washing and barley separation were observed among other macaques on the island some years later, well after the original monkey who had discovered these techniques had passed away, indicating that he had taught it to his fellows and they in turn had passed it on to the young.

Chimpanzees are more humanlike than macaques. They have a language of grunts and hoots and have been trained in captivity to understand and express themselves in a limited range of human words. In his book *Chimpanzee Politics*, de Waal describes the machinations of a group of chimps trying to achieve alpha male status in a captive colony in the Netherlands. They enter into alliances, betray one another, plead, beg, and cajole in ways that would be very familiar to Machiavelli. Chimpanzees also appear to have a sense of humor, as de Waal explains in *The Ape and the Sushi Master*:

> When guests arrive at the Field Station of the Yerkes Primate Center, near Atlanta, where I work, they usually pay a visit to my chimpanzees. Often our favorite troublemaker, a female named Georgia, hurries to the spigot to collect a mouthful of water before they arrive . . . If necessary, Georgia will wait minutes with closed lips until the visitors come near. Then there will be shrieks, laughs, jumps, and sometimes falls when she suddenly sprays them.
>
> . . . I once found myself in a similar situation with Georgia. She had taken a drink from the spigot and was sneaking up to me. I looked straight into her eyes and pointed my finger at her, warning, in Dutch, "I have seen you!" She immediately stepped back, let some of the water fall from her mouth, and swallowed the rest. I certainly do not claim that she understands Dutch, but she must have sensed that I knew what she was up to, and that I was not going to be an easy target.[33]

Georgia could apparently not just play jokes, but could feel embarrassment at being caught as well.

Examples like these are frequently cited not only to support the idea of animal rights but to denigrate human claims of uniqueness and special status. Some scientists revel in debunking traditional claims about human dignity, particularly if they are based in religion. As will be seen in the next chapter, there is still a great deal to the idea of human dignity, but the point remains that a wide variety of animals share a number of important characteristics with humans. Human beings are always making sentimental reference to their "shared humanity," but in many cases what they are referring to is their shared animality. Elephant parents, for example, appear to mourn the loss of their offspring, and become highly agitated when they discover the remains of a dead elephant. It is not too much of a stretch to imagine that a human being grieving for a lost relative or feeling dread at the sight of a corpse has something very distantly in common with the elephant (which is perhaps why we paradoxically call animal protection societies "humane" societies).

But if animals have a "right" not to suffer unduly, the nature and limits of that right depend entirely on empirical observation of what is typical for their species—that is, on a substantive judgment about their natures. To my knowledge, not even the most radical animal rights activist has ever made a case for the rights of AIDS viruses or *E. coli* bacteria, which human beings seek to destroy by the billions every day. We don't think to accord these living creatures rights because, not having nervous systems, they apparently can't suffer or be aware of their situation. We tend to accord conscious creatures greater rights in this regard because, like humans, they can anticipate suffering and have fears and hopes. A distinction of this sort might serve to distinguish the rights of a salamander from those of, say, your dog Rover—to the relief of the Watsons of the world.

But even if we accept the fact that animals have a right not to suffer unduly, there is a whole range of rights that they cannot be granted because they are not human. We would not even consider granting the right to vote, for example, to creatures that, as a group, were incapable of learning human language. Chimps can communicate in a language typical of their species, and they can master a very limited number of human words if extensively trained, but they cannot master human language and do not possess human cognition more generally. That some human beings can't master human language either actually confirms its importance to political rights: children are excluded from the right to vote because they do not as a group have the cognitive abilities of a typical adult. In all of these cases, the species-specific differences between nonhuman animals on the one hand and human beings on the other make a tremendous amount of difference to our understanding of their moral status.[34]

Blacks and women were at one time excluded from the vote in the United States on the grounds that they did not have the cognitive abilities necessary to exercise this right properly. Blacks and women can vote today, while chimps and children cannot, because of what we know empirically about the cognitive and linguistic abilities of each of these groups. Membership in one of these groups does not guarantee that one's individual characteristics will be close to the median for that group (I know a lot of individual children who would vote more wisely than their parents), but it is a good enough indicator of ability for practical purposes.

What an animal rights proponent like Peter Singer calls speciesism is thus not necessarily an ignorant and self-serving prejudice on the part of human beings, but a belief about human dignity that can be defended on the basis of an empirically grounded view of human specificity. We have broached this subject with the discussion of human cognition. But if we are to find a source of that superior human moral status that raises us all above the rest of animal creation and yet makes us equals of one another qua human beings, we need to know more about that subset of characteristics of human nature that are not just typical of our species but unique to human beings. Only then will we know what needs the greatest safeguarding against future developments in biotechnology.

## HUMAN DIGNITY

Is it, then, possible to imagine a new Natural Philosophy, continually conscious that the "natural object" produced by analysis and abstraction is not reality but only a view, and always correcting the abstraction? I hardly know what I am asking for . . . The regenerate science which I have in mind would not do even to minerals and vegetables what modern science threatens to do to man himself. When it explained it would not explain away. When it spoke of parts it would remember the whole . . . The analogy between the *Tao* of Man and the instincts of an animal species would mean for it new light cast on the unknown thing, Instinct, by the inly known reality of conscience and not a reduction of conscience to the category of Instinct. Its followers would not be free with the words *only* and *merely*. In a word, it would conquer Nature without being at the same time conquered by her and buy knowledge at a lower cost than that of life.

C. S. Lewis, *The Abolition of Man*[1]

According to the Decree by the Council of Europe on Human Cloning, "The instrumentalisation of human beings through the deliberate creation of genetically identical human beings is contrary to human dignity and thus constitutes a misuse of medicine and biology."[2] Human dignity is one of those concepts that politicians, as well as virtually everyone else in political life, like to throw around, but that almost no one can either define or explain.

Much of politics centers on the question of human dignity and the desire for recognition to which it is related. That is, human beings constantly demand that others recognize their dignity, either as individuals or as members of religious, ethnic, racial, or other kinds of groups. The struggle for recognition is not economic: what we desire is not money but that other human beings respect us in the way we think we deserve. In earlier times, rulers wanted others to recognize their superior worth as king, emperor, or lord. Today, people seek recognition of their equal status as members of formerly disrespected or devalued groups—as women, gays, Ukrainians, the handicapped, Native Americans, and the like.[3]

The demand for an equality of recognition or respect is the dominant passion of modernity, as Tocqueville noted over 170 years ago in *Democracy in America*.[4] What this means in a liberal democracy is a bit complicated. It is not necessarily that we think we are equal in all important respects, or demand that our lives be the same as everyone else's. Most people accept the fact that a Mozart or an Einstein or a Michael Jordan has talents and abilities that they don't have, and receives recognition and even monetary compensation for what he accomplishes with those talents. We accept, though we don't necessarily like, the fact that resources are distributed unequally based on what James Madison called the "different and unequal faculties of acquiring property." But we also believe that people deserve to keep what they earn and that the faculties for working and earning will not be the same for all people. We also accept the fact that we look different, come from different races and ethnicities, are of different sexes, and have different cultures.

## FACTOR X

What the demand for equality of recognition implies is that when we strip all of a person's contingent and accidental characteristics away, there remains some essential human quality underneath that is worthy of a certain minimal level of respect—call it Factor X. Skin color, looks, social class and wealth, gender, cultural background, and even one's natural talents are all accidents of birth relegated to the class of nonessential characteristics. We make decisions on whom to be-friend, whom to marry or do business with, or whom to shun at social events on the basis of these secondary characteristics. But in the po-litical realm we are required to respect people equally on the basis of their possession of Factor X. You can cook, eat, torture, enslave, or render the carcass of any creature lacking Factor X, but if you do the same thing to a human being, you are guilty of a "crime against hu-manity." We accord beings with Factor X not just human rights but, if they are adults, political rights as well—that is, the right to live in democratic political communities where their rights to speech, reli-gion, association, and political participation are respected.

The circle of beings to whom we attribute Factor X has been one of the most contested issues throughout human history. For many so-cieties, including most democratic societies in earlier periods of his-tory, Factor X belonged to a significant subset of the human race, excluding people of certain sexes, economic classes, races, and tribes and people with low intelligence, disabilities, birth defects, and the like. These societies were highly stratified, with different classes pos-sessing more or less of Factor X, and some possessing none at all. To-day, for believers in liberal equality, Factor X etches a bright red line around the whole of the human race and requires equality of respect for all of those on the inside, but attributes a lower level of dignity to those outside the boundary. Factor X is the human essence, the most basic meaning of what it is to be human. If all human beings are in fact equal in dignity, then X must be some characteristic universally possessed by them. So what is Factor X, and where does it come from?

For Christians, the answer is fairly easy: it comes from God. Man is created in the image of God, and therefore shares in some of God's sanctity, which entitles human beings to a higher level of respect than the rest of natural creation. In the words of Pope John Paul II, what this means is that "the human individual cannot be subordinated as a pure means or a pure instrument, either to the species or to society; he has value per se. He is a person. With his intellect and his will, he is capable of forming a relationship of communion, solidarity and self-giving with his peers . . . It is by virtue of his spiritual soul that the whole person possesses such dignity even in his body."[5]

Supposing one is not a Christian (or a religious believer of any sort), and doesn't accept the premise that man is created in the image of God. Is there a secular ground for believing that human beings are entitled to a special moral status or dignity? Perhaps the most famous effort to create a philosophical basis for human dignity was that of Kant, who argued that Factor X was based on the human capacity for moral choice. That is, human beings could differ in intelligence, wealth, race, and gender, but all were equally able to act according to moral law or not. Human beings had dignity because they alone had free will—not just the subjective illusion of free will but the actual ability to transcend natural determinism and the normal rules of causality. It is the existence of free will that leads to Kant's well-known conclusion that human beings are always to be treated as ends and not as means.

It would be very difficult for any believer in a materialistic account of the universe—which includes the vast majority of natural scientists—to accept the Kantian account of human dignity. The reason is that it forces them to accept a form of dualism—that there is a realm of human freedom parallel to the realm of nature that is not determined by the latter. Most natural scientists would argue that what we believe to be free will is in fact an illusion and that all human decision making can ultimately be traced back to material causes. Human beings decide to do one thing over another because one set of neurons fires rather than another, and those neuronal firings can be traced back to prior material states of the brain. The human decision-making process may be more complex than that of other animals, but there is no sharp dividing line that distinguishes human moral choice from the kinds of choices that are made by other animals. Kant himself does not offer any proof that free will exists; he says that it is simply a necessary postulate of pure practical reason about the nature of morality—hardly an argument that a hard-bitten empirical scientist would accept.

## SEIZE THE POWER

The problem posed by modern natural science goes even deeper. The very notion that there exists such a thing as a human "essence" has been under relentless attack by modern science for much of the past century and a half. One of the most fundamental assertions of Darwinism is that species do not have essences.[6] That is, while Aristotle believed in the eternity of the species (i.e., that what we have been labeling "species-typical behavior" is something unchanging), Darwin's theory maintains that this behavior changes in response to the organism's interaction with its environment. What is typical for a species represents a snapshot of the species at one particular moment of evolutionary time; what came before and what comes after will be different. Since Darwinism maintains that there is no cosmic teleology guiding the process of evolution, what seems to be the essence of a species is just an accidental by-product of a random evolutionary process.

In this perspective, what we have been calling human nature is merely the species-typical human characteristics and behavior that emerged about 100,000 years ago, during what evolutionary biologists call the "era of evolutionary adaptation"—when the precursors of modern humans were living and breeding on the African savanna. For many, this suggests that human nature has no special status as a guide to morals or values because it is historically contingent. David Hull, for example, argues,

> I do not see why the existence of human universals is all that impor-
> tant. Perhaps all and only people have opposable thumbs, use tools,
> live in true societies, or what have you. I think that such attributions
> are either false or vacuous, but even if they were true and significant,
> the distributions of these particular characters is largely a matter of
> evolutionary happenstance.[7]

The geneticist Lee Silver, trying to debunk the idea that there is a natural order that could be undermined by genetic engineering, asserts,

> Unfettered evolution is never predetermined [toward some goal], and
> not necessarily associated with progress—it is simply a response to
> unpredictable environmental changes. If the asteroid that hit our
> planet 60 million years ago had flown past instead, there would never
> have been any human beings at all. And whatever the natural order
> might be, it is not necessarily good. The smallpox virus was part of
> the natural order until it was forced into extinction by human inter-
> vention.[8]

This inability to define a natural essence doesn't bother either writer. Hull, for example, states that "I, for one, would be extremely uneasy to base something as important as human rights on such temporary contingencies [as human nature] . . . I fail to see why it matters. I fail to see, for example, why we must all be essentially the same to have rights."[9] Silver, for his part, pooh-poohs fears about genetic engineering on the part of those with religious convictions or those who believe in a natural order. In the future, man will no longer be a slave to his genes, but their master:

> Why not seize this power? Why not control what has been left to chance in the past? Indeed, we control all other aspects of our children's lives and identities through powerful social and environmental influences and, in some cases, with the use of powerful drugs like Ritalin and Prozac. On what basis can we reject positive genetic influences on a person's essence when we accept the rights of parents to benefit their children in every other way?[10]

Why not seize this power, indeed?

Well, let us begin by considering what the consequences of the abandonment of the idea that there is a Factor X, or human essence, that unites all human beings would be for the cherished idea of universal human equality—an idea to which virtually all of the debunkers of the idea of human essences are invariably committed. Hull is right that we don't all need to be the same in order to have rights—but we need to be the same in some one critical respect in order to have *equal* rights. He for one is very concerned that basing human rights on human nature will stigmatize homosexuals, because their sexual orientation differs from the heterosexual norm. But the only basis on which anyone can make an argument in favor of equal rights for gays is to argue that whatever their sexual orientation, *they are people too* in some other respect that is more essential than their sexuality. If you cannot find this common other ground, then there is no reason not to discriminate against them, because in fact they are different creatures from everyone else.

Similarly, Lee Silver, who is so eager to take up the power of genetic engineering to "improve" people, is nonetheless horrified at the possibility that it could be used to create a class of genetically superior people. He paints a scenario in which a class called the GenRich steadily improve the cognitive abilities of their children to the point that they break off from the rest of the human race to form a separate species.

Silver is not horrified by much else that technology may bring us by way of unnatural reproduction—for example, two lesbians producing genetic offspring, or eggs taken from an unborn female fetus to produce a child whose mother had never been born. He dismisses the moral concerns of virtually every religion or traditional moral system with regard to future genetic engineering but draws the line at what he perceives as threats to human equality. He does not seem to understand that, given his premises, there are no possible grounds on which he can object to the GenRich, or the fact that they might assign themselves rights superior to those of the GenPoor. Since there is no stable essence common to all human beings, or rather because that essence is variable and subject to human manipulation, why not create a race born with metaphorical saddles on their backs, and another with boots and spurs to ride them? Why not seize *that* power as well?

The bioethicist Peter Singer, whose appointment to Princeton University caused great controversy because of his advocacy of infanticide and euthanasia under certain circumstances, is simply more consistent than most people on the consequences of abandoning the concept of human dignity. Singer is an unabashed utilitarian: he believes that the single relevant standard for ethics is to minimize suffering in the aggregate for all creatures. Human beings are part of a continuum of life and have no special status in his avowedly Darwinian worldview. This leads him to two perfectly logical conclusions: the need for animal rights, since animals can experience pain and suffering as well as humans, and the downgrading of the rights of infants and elderly people who lack certain key traits, like self-awareness, that would allow them to anticipate pain. The rights of certain animals, in his view, deserve greater respect than those of certain human beings.

But Singer is not nearly forthright enough in following these premises through to their logical conclusion, since he remains a committed egalitarian. What he does not explain is why the relief of suffering should remain the only moral good. As usual, the philosopher Friedrich Nietzsche was much more clear-eyed than anyone else in understanding the consequences of modern natural science and the abandonment of the concept of human dignity. Nietzsche had the great insight to see that, on the one hand, once the clear red line around the whole of humanity could no longer be drawn, the way would be paved for a return to a much more hierarchical ordering of society. If there is a continuum of gradations between human and nonhuman, there is a continuum within the type human as well. This would inevitably mean the liberation of the strong from the constraints that a belief in either God or Nature had placed on them. On the other hand, it would lead the rest of mankind to demand health and safety as the only possible goods, since all the higher goals that had once been set for them were now debunked. In the words of Nietzsche's Zarathustra, "One has one's little pleasure for the day and one's little pleasure for the night: but one has a regard for health. 'We have invented happiness,' say the last men, and they blink."[11] Indeed, both the return of hierarchy and the egalitarian demand for health, safety, and relief of suffering might all go hand in hand if the rulers of the future could provide the masses with enough of the "little poisons" they demanded.

It has always struck me that one hundred years after Nietzsche's death, we are much less far down the road to either the superman or the last man than he predicted. Nietzsche once castigated John Stuart Mill as a "flathead" for believing that one could have a semblance of Christian morality in the absence of belief in a Christian God. And yet, in a Europe and an America that have become secularized over the past two generations, we see a lingering belief in the concept of human dignity, which is by now completely cut off from its religious roots. And not just lingering: the idea that one could exclude any group of people on the basis of race, gender, disability, or virtually any other characteristic from the charmed circle of those deserving recognition for human dignity is the one thing that will bring total obloquy on the head of any politician who proposes it. In the words of the philosopher Charles Taylor, "We believe it would be utterly wrong and unfounded to draw the boundaries any narrower than around the whole human race," and should anyone try to do so, "we should immediately ask what distinguished those within from those left out."[12] The idea of the equality of human dignity, deracinated from its Christian or Kantian origins, is held as a matter of religious dogma by the most materialist of natural scientists. The continuing arguments over the moral status of the unborn (about which more later) constitute the only exception to this general rule.

The reasons for the persistence of the idea of the equality of human dignity are complex. Partly it is a matter of the force of habit and what Max Weber once called the "ghost of dead religious beliefs" that continue to haunt us. Partly it is the product of historical accident: the last important political movement to explicitly deny the premise of universal human dignity was Nazism, and the horrifying consequences of the Nazis' racial and eugenic policies were sufficient to inoculate those who experienced them for the next couple of generations.

But another important reason for the persistence of the idea of the universality of human dignity has to do with what we might call the nature of nature itself. Many of the grounds on which certain groups were historically denied their share of human dignity were proven to be simply a matter of prejudice, or else based on cultural and environmental conditions that could be changed. The notions that women were too irrational or emotional to participate in politics, and that immigrants from southern Europe had smaller head sizes and were less intelligent than those from northern Europe, were overturned on the basis of sound, empirical science. That moral order did not completely break down in the West in the wake of the destruction of consensus over traditional religious values should not surprise us either, because moral order comes from within human nature itself and is not something that has to be imposed on human nature by culture.[13]

All of this could change under the impact of future biotechnology. The most clear and present danger is that the large genetic variations between individuals will narrow and become clustered within certain distinct social groups. Today, the "genetic lottery" guarantees that the son or daughter of a rich and successful parent will not necessarily inherit the talents and abilities that created conditions conducive to the parent's success. Of course, there has always been a degree of genetic selection: assortative mating means that successful people will tend to marry each other and, to the extent that their success is genetically based, will pass on to their children better life opportunities. But in the future, the full weight of modern technology can be put in the service of optimizing the kinds of genes that are passed on to one's offspring. This means that social elites may not just pass on social advantages but embed them genetically as well. This may one day include not only characteristics like intelligence and beauty, but behavioral traits like diligence, competitiveness, and the like.

The genetic lottery is judged as inherently unfair by many because it condemns certain people to lesser intelligence, or bad looks, or disabilities of one sort or another. But in another sense it is profoundly egalitarian, since everyone, regardless of social class, race, or ethnicity, has to play in it. The wealthiest man can and often does have a good-for-nothing son; hence the saying "Shirtsleeves to shirtsleeves in three generations." When the lottery is replaced by choice, we open up a new avenue along which human beings can compete, one that threatens to increase the disparity between the top and bottom of the social hierarchy.

What the emergence of a genetic overclass will do to the idea of universal human dignity is something worth pondering. Today, many bright and successful young people believe that they owe their success to accidents of birth and upbringing but for which their lives might have taken a very different course. They feel themselves, in other words, to be lucky, and they are capable of feeling sympathy for people who are less lucky than they. But to the extent that they become "children of choice" who have been genetically selected by their parents for certain characteristics, they may come to believe increasingly that their success is a matter not just of luck but of good choices and planning on the part of their parents, and hence something deserved. They will look, think, act, and perhaps even feel differently from those who were not similarly chosen, and may come in time to think of themselves as different kinds of creatures. They may, in short, feel themselves to be aristocrats, and unlike aristocrats of old, their claim to better birth will be rooted in nature and not convention.

Aristotle's discussion of slavery in Book I of the *Politics* is instructive on this score. It is often condemned as a justification of Greek slavery, but in fact the discussion is far more sophisticated and is relevant to our thinking about genetic classes. Aristotle makes a distinction between conventional and natural slavery.[14] He argues that slavery would be justified by nature if it were the case that there were people with naturally slavish natures. It is not clear from his discussion that he believes such people exist: most actual slavery is conventional—that is, it is the result of victory in war or force, or based on the wrong opinion that barbarians as a class should be slaves of Greeks.[15] The noble-born think their nobility comes from nature rather than acquired virtue and that they can pass it on to their children. But, Aristotle notes, nature is "frequently unable to bring this about."[16] So why not, as Lee Silver suggests, "seize this power" to give children genetic advantages and correct the defect of natural equality?

The possibility that biotechnology will permit the emergence of new genetic classes has been frequently noted and condemned by those who have speculated about the future.[17] But the opposite possibility also seems to be entirely plausible—that there will be an impetus toward a much more genetically egalitarian society. For it seems highly unlikely that people in modern democratic societies will sit around complacently if they see elites embedding their advantages genetically in their children.

Indeed, this is one of the few things in a politics of the future that people are likely to rouse themselves to fight over. By this I mean not just fighting metaphorically, in the sense of shouting matches among talking heads on TV and debates in Congress, but actually picking up guns and bombs and using them on other people. There are very few domestic political issues today in our rich, self-satisfied liberal democracies that can cause people to get terribly upset, but the specter of rising genetic inequality may well get people off their couches and into the streets.

If people get upset enough about genetic inequality, there will be two alternative courses of action. The first and most sensible would simply be to forbid the use of biotechnology to enhance human characteristics and decline to compete in this dimension. But the notion of enhancement may become too powerfully attractive to forgo, or it may prove difficult to enforce a rule preventing people from enhancing their children, or the courts may declare they have a right to do so.

At this point a second possibility opens up, which is to use that same technology to raise up the bottom.[18]

This is the only scenario in which it is plausible that we will see a liberal democracy of the future get back into the business of state-sponsored eugenics. The bad old form of eugenics discriminated against the disabled and less intelligent by forbidding them to have children. In the future, it may be possible to breed children who are more intelligent, more healthy, more "normal." Raising the bottom is something that can only be accomplished through the intervention of the state. Genetic enhancement technology is likely to be expensive and involve some risk, but even if it were relatively cheap and safe, people who are poor and lacking in education would still fail to take advantage of it. So the bright red line of universal human dignity will have to be reinforced by allowing the state to make sure that no one falls outside it.

The politics of breeding future human beings will be very complex. Up to now, the Left has on the whole been opposed to cloning, genetic engineering, and similar biotechnologies for a number of reasons, including traditional humanism, environmental concerns, suspicion of technology and of the corporations that produce it, and fear of eugenics. The Left has historically sought to play down the importance of heredity in favor of social factors in explaining human outcomes. For people on the Left to come around and support genetic engineering for the disadvantaged, they would first have to admit that genes are important in determining intelligence and other types of social outcomes in the first place.

The Left has been more hostile to biotechnology in Europe than in North America. Much of this hostility is driven by the stronger environmental movements there, which have led the campaign, for example, against genetically modified foods. (Whether certain forms of radical environmentalism will translate into hostility to human biotechnology remains to be seen. Some environmentalists see themselves defending nature from human beings, and seem to be more concerned with threats to nonhuman than to human nature.) The Germans in particular remain very sensitive to anything that smacks of eugenics. The philosopher Peter Sloterdijk raised a storm of protest in 1999 when he suggested that it will soon be impossible for people to refuse the power of selection that biotechnology provides them, and that the questions of breeding something "beyond" man that were raised by Nietzsche and Plato could no longer be ignored.[19] He was condemned by the sociologist Jürgen Habermas, among others, who in other contexts has also come out against human cloning.[20]

On the other hand, there are some on the Left who have begun to make the case for genetic engineering.[21] John Rawls argued in *A Theory of Justice* that the unequal distribution of natural talents was inherently unfair. A Rawlsian should therefore want to make use of biotechnology to equalize life chances by breeding the bottom up, assuming that prudential considerations concerning safety, cost, and the like would be settled. Ronald Dworkin has laid out a case for the right of parents to genetically engineer their children based on a broader concern to protect autonomy,[22] and Laurence Tribe has suggested that a ban on cloning would be wrong because it might create discrimination against children who were cloned in spite of the ban.[23]

It is impossible to know which of these two radically different scenarios—one of growing genetic inequality, the other of growing genetic equality—is more likely to come to pass. But once the technological possibility for biomedical enhancement is realized, it is hard to see how growing genetic inequality would fail to become one of the chief controversies of twenty-first-century politics.

## HUMAN DIGNITY REDUX

Denial of the concept of human dignity—that is, of the idea that there is something unique about the human race that entitles every member of the species to a higher moral status than the rest of the natural world—leads us down a very perilous path. We may be compelled ultimately to take this path, but we should do so only with our eyes open. Nietzsche is a much better guide to what lies down that road than the legions of bioethicists and casual academic Darwinians that today are prone to give us moral advice on this subject.

To avoid following that road, we need to take another look at the notion of human dignity, and ask whether there is a way to defend the concept against its detractors that is fully compatible with modern natural science but that also does justice to the full meaning of human specificity. I believe that there is.

In contrast to a number of conservative Protestant denominations that continue to hold a brief for creationism, the Catholic Church by the end of the twentieth century had come to terms with the theory of evolution. In his 1996 message to the Pontifical Academy of Sciences, Pope John Paul II corrected the encyclical *Humani generis* of Pius XII, which maintained that Darwinian evolution was a serious hypothesis but one that remained unproven. The pope stated, "Today, almost half a century after the publication of the Encyclical, new knowledge has led to the recognition of the theory of evolution as more than a hypothesis. It is indeed remarkable that this theory has been progressively accepted by researchers, following a series of discoveries in various fields of knowledge. The convergence, neither sought nor fabricated, of the results of work that was conducted independently is in itself a significant argument in favor of this theory."[24]

But the pope went on to say that while the church can accept the view that man is descended from nonhuman animals, there is an "ontological leap" that occurs somewhere in this evolutionary process.[25] The human soul is something directly created by God: consequently, "theories of evolution which, in accordance with the philosophies inspiring them, consider the mind as emerging from the forces of living nature, or as a mere epiphenomenon of this matter, are incompatible with the truth about man." The pope continued, "Nor are they able to ground the dignity of the person."

The pope was saying, in other words, that at some point in the 5 million years between man's chimplike forebears and the emergence of modern human beings, a human soul was inserted into us in a way that remains mysterious. Modern natural science can uncover the time line of this process and explicate its material correlates, but it has not fully explained either what the soul is or how it came to be. The church has obviously learned a great deal from modern natural science in the past two centuries and has adjusted its doctrines accordingly. But while many natural scientists would scoff at the idea that they have anything to learn from the church, the pope has pointed to a real weakness in the current state of evolutionary theory, which scientists would do well to ponder. Modern natural science has explained a great deal less about what it means to be human than many scientists think it has.

### Parts and Wholes

Many contemporary Darwinians believe that they have demystified the problem of how human beings came to be human through the classical reductionist methods of modern natural science. That is, any higher-order behavior or characteristic, such as language or aggression, can be traced back through the firing of neurons to the biochemical substrate of the brain, which in turn can be understood in terms of the simpler organic compounds of which it is composed. The brain arrived at its present state through a series of incremental evolutionary changes that were driven by random variation, and a process of natural selection by which the requirements of the surrounding environment selected for certain mental characteristics. Every human characteristic can thus be traced back to a prior material cause. If, for example, we today love to listen to Mozart or Beethoven, it is because we have auditory systems that were evolved, in the environment of evolutionary adaptation, to discriminate between certain kinds of sounds that were necessary perhaps to warn us against predators or to help us on a hunt.[26]

The problem with this kind of thinking is not that it is necessarily false but that it is insufficient to explain many of the most salient and unique human traits. The problem lies in the methodology of reductionism itself for understanding complex systems, and particularly biological ones.

Reductionism constitutes, of course, one of the foundations of modern natural science and is responsible for many of its greatest triumphs. You see before you two apparently different substances, the graphite in your pencil lead and the diamond in your engagement ring, and you might be tempted to believe that they were essentially different substances. But reductionist chemistry has taught us that in fact they are both composed of the same simpler substance, carbon, and that the apparent differences are not ones of essence but merely of the way the carbon atoms are bonded. Reductionist physics has been busy over the past century tracing atoms back to subatomic particles and thence back to an even more reduced set of basic forces of nature.

But what is appropriate for domains in physics, like celestial mechanics and fluid dynamics, is not necessarily appropriate for the study of objects at the opposite end of the complexity scale, like most biological systems, because the behavior of complex systems cannot be predicted by simply aggregating or scaling up the behavior of the parts that constitute them.* The distinctive and easily recognizable behavior of a flock of birds or a swarm of bees, for example, is the product of the interaction of individual birds or bees following relatively simple behavioral rules (fly next to a partner, avoid obstacles, and so on), none of which encompasses or defines the behavior of the flock or swarm as a whole. Rather, the group behavior "emerges" as a result of the interaction of the individuals that make it up. In many cases, the relationship between parts and wholes is nonlinear: that is, increasing input A increases output B up to a certain point, whereupon it creates a qualitatively different and unexpected output C. This is true even of relatively simple chemicals like water: $H_2O$ undergoes a phase transition from liquid to solid at 32 degrees Fahrenheit, something that one would not necessarily predict on the basis of knowledge of its chemical composition.

That the behavior of complex wholes cannot be understood as the aggregated behavior of their parts has been understood in the natural sciences for some time now,[27] and has led to the development of the field of so-called nonlinear or "complex adaptive" systems, which try to model the emergence of complexity. This approach is, in a way, the opposite of reductionism: it shows that while wholes can be traced back to their simpler antecedent parts, there is no simple predictive model that allows us to move from the parts to the emergent behaviors of the wholes. Being nonlinear, they may be extremely sensitive to small differences in starting conditions and thus may appear chaotic even when their behavior is completely deterministic.

This means that the behavior of complex systems is much more difficult to understand than the founders of reductionist science once

---

*The determinism of classical Newtonian mechanics is based in large measure on the parallelogram rule, which says that the effects of two forces acting on a body can be summed as if each were acting independently of the other. Newton shows that this rule works for celestial bodies like planets and stars, and assumes that it will also work for other natural objects, like animals.

believed. The eighteenth-century astronomer Laplace once said that he could precisely predict the future of the universe on the basis of Newtonian mechanics, if he could know the mass and motion of the universe's constituent parts.[28] No scientist could make this claim to-day—not just because of the inherent uncertainties introduced by quantum mechanics but also because there exists no reliable method-ology for predicting the behavior of complex systems.[29] In the words of Arthur Peacocke, "The concepts and theories . . . that constitute the content of the sciences focusing on the more complex levels are often (not always) logically not reducible to those operative in the sci-ences that focus on their components."[30] There is a hierarchy of levels of complexity in the sciences, with human beings and human behav-ior occupying a place at the uppermost level.

Each level can give us some insight into the levels above it, but understanding the lower levels does not allow one to fully understand the higher levels' emergent properties. Researchers in the area of complex adaptive systems have created so-called agent-based models of complex systems, and have applied them in a wide variety of areas, from cell biology to fighting a war to distributing natural gas. It re-mains to be seen, however, whether this approach constitutes a sin-gle, coherent methodology applicable to all complex systems.[31] Such models may tell us only that certain systems will remain inherently chaotic and unpredictable, or that prediction rests on a precise knowledge of initial conditions that is unavailable to us. The higher level must thus be understood with a methodology appropriate to its degree of complexity.

We can illustrate the problematic relationship of parts to wholes by reference to one unique domain of human behavior, politics.[32] Aristotle states that man is a political animal by nature. If one were to try to build a case for human dignity based on human specificity, the capability of engaging in politics would certainly constitute one im-portant component of human uniqueness. Yet the idea of our unique-ness in this regard has been challenged. As noted in Chapter 8, chimpanzees and other primates engage in something that looks un-cannily like human politics as they struggle and connive to achieve al-pha male status. They appear, moreover, to feel the political emotions of pride and shame as they interact with other members of their group. Their political behavior can also apparently be transmitted through nongenetic means, so that political culture would not seem to be the exclusive preserve of human beings.[33] Some observers gleefully cite examples like this to deflate human feelings of self-importance relative to other species.

But to confuse human politics with the social behavior of any other species is to mistake parts for wholes. Only human beings can formulate, debate, and modify abstract rules of justice. When Aristotle asserted that man is a political animal by nature, he meant this only in the sense that politics is a potentiality that emerges over time.[34] He notes that human politics did not begin until the first law-giver established a state and promulgated laws, an event that was of great benefit to mankind but that was contingent on historical developments. This accords with what we know today about the emergence of the state, which took place in parts of the world like Egypt and Babylonia perhaps 10,000 years ago and was most likely related to the development of agriculture. For tens of thousands of years before that, human beings lived in stateless hunter-gatherer societies in which the largest group numbered no more than 50 to 100 individuals, most of them related by kinship.[35] So in a certain sense, while human sociability is obviously natural, it is not clear that humans are political animals by nature.

But Aristotle insists that politics is natural to man despite the fact that it did not exist at all in early periods of human history. He argues that it is human language that allows human beings to formulate laws and abstract principles of justice that are necessary to the creation of a state and of political order. Ethologists have noted that many other species communicate with sounds, and that chimpanzees and other animals can learn human language to a limited extent. But no other species has *human* language—that is, the ability to formulate and communicate abstract principles of action. It is only when these two natural characteristics, human sociability and human language, come together that human politics emerges. Human language obviously evolved to promote sociability, but it is very unlikely that there were evolutionary forces shaping it to become an enabler of politics. It was rather like one of Stephen Jay Gould's spandrels,* something

---

*A spandrel is an architectural feature that emerges, unplanned by the architect, from the intersection of a dome and the walls that support it.

that evolved for one reason but that found another key purpose when combined in a human whole.[36] Human politics, though natural in an emergent sense, is not reducible to either animal sociability or animal language, which were its precursors.

### Consciousness

The area in which the inability of a reductionist materialist science to explain observable phenomena is most glaringly evident is the question of human consciousness. By consciousness I mean subjective mental states: not just the thoughts and images that appear to you as you are thinking or reading this page, but also the sensations, feelings, and emotions that you experience as part of everyday life.

There has been a huge amount of research and theorizing about consciousness over the past two generations, coming in equal measure from the neurosciences and from studies in computer and artificial intelligence (AI). Particularly in the latter field there are many enthusiasts who are convinced that with more powerful computers and new approaches to computing, such as neural networks, we are on the verge of a breakthrough in which mechanical computers will achieve consciousness. There have been conferences and earnest discussions devoted to the question of whether it would be moral to turn off such a machine if and when this breakthrough occurs, and whether we would need to assign rights to conscious machines.

The fact of the matter is that we are nowhere close to a breakthrough; consciousness remains as stubbornly mysterious as it ever was. The problem with the current state of thinking begins with the traditional philosophical problem of the ontological status of consciousness. Subjective mental states, while produced by material biological processes, appear to be of a very different, nonmaterial order from other phenomena. The fear of dualism—that is, the doctrine that there are two essential types of being, material and mental—is so strong among researchers in this field that it has led them to palpably ridiculous conclusions. In the words of the philosopher John Searle,

> Seen from the perspective of the last fifty years, the philosophy of
> mind, as well as cognitive science and certain branches of psychol-
> ogy, present a very curious spectacle. The most striking feature is how
> much of mainstream philosophy of mind of the past fifty years seems

obviously false . . . in the philosophy of mind, obvious facts about the mental, such as that we all really do have subjective conscious mental states and that these are not eliminable in favor of anything else, are routinely denied by many, perhaps most, of the advanced thinkers in the subject.[37]

An example of a patently false understanding of consciousness comes from one of the leading experts in the field, Daniel Dennett, whose book *Consciousness Explained* finally comes to the following definition of consciousness: "Human consciousness is *itself* a huge complex of memes (or more exactly, meme-effects in brains) that can best be understood as the operation of a '*von Neumannesque*' virtual machine *implemented* in the *parallel architecture* of a brain that was not designed for any such activities."[38] A naive reader may be excused for thinking that this kind of statement doesn't do much at all to advance our understanding of consciousness. Dennett is saying in effect that human consciousness is simply the by-product of the operations of a certain type of computer, and if we think that there is more to it than that, we have a mistakenly old-fashioned view of what consciousness is. As Searle says of this approach, it works only by denying the existence of what you and I and everyone else understand consciousness to be (that is, subjective feelings).[39]

Similarly, many of the researchers in the field of artificial intelligence sidestep the question of consciousness by in effect changing the subject. They assume that the brain is simply a highly complex type of organic computer that can be identified by its external characteristics. The well-known Turing test asserts that if a machine can perform a cognitive task such as carrying on a conversation in a way that from the outside is indistinguishable from similar activities carried out by a human being, then it is indistinguishable on the inside as well. Why this should be an adequate test of human mentality is a mystery, for the machine will obviously not have any subjective awareness of what it is doing, or feelings about its activities.* This doesn't

---

*Searle's critique of this approach is contained in his "Chinese room" puzzle, which raises the question of whether a computer could be said to understand Chinese any more than a non-Chinese-speaking individual locked in a room who received instructions on how to manipulate a series of symbols in Chinese. See Searle (1997), p. 11.

prevent such authors as Hans Moravec[40] and Ray Kurzweil[41] from predicting that machines, once they reach a requisite level of complexity, will possess human attributes like consciousness as well.[42] If they are right, this will have important consequences for our notions of human dignity, because it will have been conclusively proven that human beings are essentially nothing more than complicated machines that can be made out of silicon and transistors as easily as carbon and neurons.

The likelihood that this will happen seems very remote, however, not so much because machines will never duplicate human intelligence—I suspect they will probably be able to come very close in this regard—but rather because it is impossible to see how they will come to acquire human emotions. It is the stuff of science fiction for an android, robot, or computer to suddenly start experiencing emotions like fear, hope, even sexual desire, but no one has come remotely close to positing how this might come about. The problem is not simply that, like the rest of consciousness, no one understands what emotions are ontologically; no one understands why they came to exist in human biology.

There are of course functional reasons for feelings like pain and pleasure. If we didn't find sex pleasurable we wouldn't reproduce, and if we didn't feel pain from fire we would be burning ourselves constantly. But state-of-the-art thinking in cognitive science maintains that the particular subjective form that the emotions take is not necessary to their function. It is perfectly possible, for example, to design a robot with heat sensors in its fingers connected to an actuator that would pull the robot's hand away from a fire. The robot could keep itself from being burned without having any subjective sense of pain, and it could make decisions on which objectives to fulfill and which activities to avoid on the basis of a mechanical computation of the inputs of different electrical impulses. A Turing test would say it was a human being in its behavior, but it would actually be devoid of the most important quality of a human being, feelings. The actual subjective form that emotions take are today seen in evolutionary biology and in cognitive science as no more than epiphenomenal to their underlying function; there are no obvious reasons this form should have been selected for in the course of evolutionary history.[43]

As Robert Wright points out, this leads to the very bizarre outcome that what is most important to us as human beings has no apparent purpose in the material scheme of things by which we became human.[44] For it is the distinctive human gamut of emotions that produces human purposes, goals, objectives, wants, needs, desires, fears, aversions, and the like and hence is the source of human values. While many would list human reason and human moral choice as the most important unique human characteristics that give our species dignity, I would argue that possession of the full human emotional gamut is at least as important, if not more so.

The political theorist Robert McShea demonstrates the importance of human emotions to our commonsense understanding of what it means to be human by asking us to perform the following thought experiment.[45] Suppose you met two creatures on a desert island, both of which had the rational capacity of a human being and hence the ability to carry on a conversation. One had the physical form of a lion but the emotions of a human being, while the other had the physical form of a human being but the emotional characteristics of a lion. Which creature would you feel more comfortable with, which creature would you be more likely to befriend or enter into a moral relationship with? The answer, as countless children's books with sympathetic talking lions suggest, is the lion, because species-typical human emotions are more critical to our sense of our own humanness than either our reason or our physical appearance. The coolly analytical Mr. Spock in the TV series *Star Trek* appears at times more likable than the emotional Mr. Scott only because we suspect that somewhere beneath his rational exterior lurk deeply buried human feelings. Certainly many of the female characters he encountered in the series hoped they could rouse something more than robotic responses from him.

On the other hand, we would regard a Mr. Spock who was truly devoid of any feelings as a psychopath and a monster. If he offered us a benefit, we might accept it but would feel no gratitude because we would know it was the product of rational calculation on his part and not goodwill. If we double-crossed him, we would feel no guilt, because we know that he cannot himself entertain feelings of anger or of having been betrayed. And if circumstances forced us to kill him to save ourselves, or to sacrifice his life in a hostage situation, we would feel no more regret than if we lost any other valuable asset, like a car or a teleporter.[46] Even though we might want to cooperate with this Mr. Spock, we would not regard him as a moral agent entitled to the respect that human beings command. The computer geeks in AI labs who think of themselves as nothing more than complex computer programs and want to download themselves into a computer should worry, since no one would care if they were turned off for good.

So there is a great deal that comes together under the rubric of consciousness that helps define human specificity and hence human dignity, which nonetheless cannot currently be fully explicated by modern natural science. It is not sufficient to argue that some other animals are conscious, or have culture, or have language, for their consciousness does not combine human reason, human language, human moral choice, and human emotions in ways that are capable of producing human politics, human art, or human religion. All of the nonhuman precursors of these human traits that existed in evolutionary history, and all of the material causes and preconditions for their emergence, collectively add up to much less than the human whole. Jared Diamond in his book *The Third Chimpanzee* notes the fact that the chimpanzee and human genomes overlap by more than 98 percent, implying that the differences between the two species are relatively trivial.[47] But for an emergent complex system, small differences can lead to enormous qualitative changes. It is a bit like saying there is no significant difference between ice and liquid water because they differ in temperature by only 1 degree.

Thus one does not have to agree with the pope that God directly inserted a human soul in the course of evolutionary history to acknowledge with him that there was a very important qualitative, if not ontological, leap that occurred at some point in this process. It is this leap from parts to a whole that ultimately has to constitute the basis for human dignity, a concept one can believe in even if one does not begin from the pope's religious premises.

What this whole is and how it came to be remain, in Searle's word, "mysterious." None of the branches of modern natural science that have tried to address this question have done more than scratch the surface, despite the belief of many scientists that they have demystified the entire process. It is common now for many AI researchers to say that consciousness is an "emergent property" of a certain kind of complex computer. But this is no more than an unproven hypothesis based on an analogy with other complex systems. No one has ever seen consciousness emerge under experimental conditions, or even posited a theory as to how this might come about. It would be surprising if the process of "emergence" didn't play an important part in explaining how humans came to be human, but whether that is all there is to the story is something we do not at present know.

This is not to say that the demystification by science will never happen. Searle himself believes that consciousness is a biological property of the brain much like the firing of neurons or the production of neurotransmitters and that biology will someday be able to explain how organic tissue can produce it. He argues that our present problems in understanding consciousness do not require us to adopt a dualistic ontology or abandon the scientific framework of material causation. The problem of how consciousness arose does not require recourse to the direct intervention of God.

It does not, on the other hand, rule it out, either.

## WHAT TO FIGHT FOR

If what gives us dignity and a moral status higher than that of other living creatures is related to the fact that we are complex wholes rather than the sum of simple parts, then it is clear that there is no simple answer to the question, What is Factor X? That is, Factor X cannot be reduced to the possession of moral choice, or reason, or language, or sociability, or sentience, or emotions, or consciousness, or any other quality that has been put forth as a ground for human dignity. It is all of these qualities coming together in a human whole that make up Factor X. Every member of the human species possesses a genetic endowment that allows him or her to become a whole human being, an endowment that distinguishes a human in essence from other types of creatures.

A moment's reflection will show that none of the key qualities that contribute to human dignity can exist in the absence of the others.

Human reason, for example, is not that of a computer; it is pervaded by emotions, and its functioning is in fact facilitated by the latter.[48] Moral choice cannot exist without reason, needless to say, but it is also grounded in feelings such as pride, anger, shame, and sympathy.[49] Human consciousness is not just individual preferences and instrumental reason, but is shaped intersubjectively by other consciousnesses and their moral evaluations. We are social and political animals not merely because we are capable of game-theoretic reason, but because we are endowed with certain social emotions. Human sentience is not that of a pig or a horse, because it is coupled with human memory and reason.

This protracted discussion of human dignity is intended to answer the following question: What is it that we want to protect from any future advances in biotechnology? The answer is, we want to protect the full range of our complex, evolved natures against attempts at self-modification. We do not want to disrupt either the unity or the continuity of human nature, and thereby the human rights that are based on it.

If Factor X is related to our very complexity and the complex interactions of uniquely human characteristics like moral choice, reason, and a broad emotional gamut, it is reasonable to ask how and why biotechnology would seek to make us less complex. The answer lies in the constant pressure that exists to reduce the ends of biomedicine to utilitarian ones—that is, the attempt to reduce a complex diversity of natural ends and purposes to just a few simple categories like pain and pleasure, or autonomy. There is in particular a constant predisposition to allow the relief of pain and suffering to automatically trump all other human purposes and objectives. For this will be the constant trade-off that biotechnology will pose: we can cure this disease, or prolong this person's life, or make this child more tractable, at the expense of some ineffable human quality like genius, or ambition, or sheer diversity.

That aspect of our complex natures most under threat has to do with our emotional gamut. We will be constantly tempted to think that we understand what "good" and "bad" emotions are, and that we can do nature one better by suppressing the latter, by trying to make people less aggressive, more sociable, more compliant, less depressed.

The utilitarian goal of minimizing suffering is itself very problematic. No one can make a brief in favor of pain and suffering, but the fact of the matter is that what we consider to be the highest and most admirable human qualities, both in ourselves and in others, are often related to the way that we react to, confront, overcome, and frequently succumb to pain, suffering, and death. In the absence of these human evils there would be no sympathy, compassion, courage, heroism, solidarity, or strength of character.* A person who has not confronted suffering or death has no depth. Our ability to experience these emotions is what connects us potentially to all other human beings, both living and dead.

Many scientists and researchers would say that we don't need to worry about fencing off human nature, however defined, from biotechnology, because we are a very long way from being able to modify it, and may never achieve the capability. They may be right: human germ-line engineering and the use of recombinant DNA technology on humans is probably much further off than many people assume, though human cloning is not.

But our ability to manipulate human behavior is not dependent on the development of genetic engineering. Virtually everything we can anticipate being able to do through genetic engineering we will most likely be able to do much sooner through neuropharmacology. And we will face large demographic changes in the populations that find new biomedical technologies available to them, not only in terms of age and sex distributions, but in terms of the quality of life of important population groups.

The widespread and rapidly growing use of drugs like Ritalin and Prozac demonstrates just how eager we are to make use of technology to alter ourselves. If one of the key constituents of our nature, something on which we base our notions of dignity, has to do with the gamut of normal emotions shared by human beings, then we are *already* trying to narrow the range for the utilitarian ends of health and convenience.

Psychotropic drugs do not alter the germ line or produce heritable

---

*The Greek root of *sympathy* and the Latin root of *compassion* both refer to the ability to feel another person's pain and suffering.

effects in the way that genetic engineering someday might. But they already raise important issues about the meaning of human dignity and are a harbinger of things to come.

### When Do We Become Human?

In the near term, the big ethical controversies raised by biotechnology will not be threats to the dignity of normal adult human beings but rather to those who possess something less than the full complement of capabilities that we have defined as characterizing human specificity. The largest group of beings in this category are the unborn, but it could also include infants, the terminally sick, elderly people with debilitating diseases, and the disabled.

This issue has already come up with regard to stem cell research and cloning. Embryonic stem cell research requires the deliberate destruction of embryos, while so-called therapeutic cloning requires not just their destruction but their deliberate creation for research purposes prior to destruction. (As bioethicist Leon Kass notes, therapeutic cloning is not therapeutic for the embryo.) Both activities have been strongly condemned by those who believe that life begins at conception and that embryos have full moral status as human beings.

I do not want to rehearse the whole history of the abortion debate and the hotly contested question of when life begins. I personally do not begin with religious convictions on this issue and admit to considerable confusion in trying to think through its rights and wrongs. The question here is, What does the natural-rights approach to human dignity outlined here suggest about the moral status of the unborn, the disabled, and so on? I'm not sure it produces a definitive answer, but it can at least help us frame an answer to the question.

At first blush, a natural-rights doctrine that bases human dignity on the fact that the human species possesses certain unique characteristics would appear to allow a gradation of rights depending on the degree to which any individual member of that species shares in those characteristics. An elderly person with Alzheimer's, for example, has lost the normal adult ability to reason, and therefore that part of his dignity that would permit him to participate in politics by voting or running for office. Reason, moral choice, and possession of the species-typical emotional gamut are things that are shared by virtually all human beings and therefore serve as a basis for universal equality, but individuals possess these traits in greater or lesser amounts: some are more reasonable, have stronger consciences or more sensitive emotions than others. At one extreme, minute distinctions could be made between individuals based on the degree to which they possess these basic human qualities, with differentiated rights assigned to them on that basis. This has happened before in history; it is called natural aristocracy. The hierarchical system it implies is one of the reasons people have become suspicious of the very concept of natural rights.

There is a strong prudential reason for not being too hierarchical in the assignment of political rights, however. There is, in the first place, no consensus on a precise definition of that list of essential human characteristics that qualify an individual for rights. More important, judgments about the degree to which a given individual possesses one or another of these qualities are very difficult to make, and usually suspect, because the person making the judgment is seldom a disinterested party. Most real-world aristocracies have been conventional rather than natural, with the aristocrats assigning themselves rights that they claimed were natural but that were actually based on force or convention. It is therefore appropriate to approach the question of who qualifies for rights with some liberality.

Nonetheless, every contemporary liberal democracy does in fact differentiate rights based on the degree to which individuals or categories of individuals share in certain species-typical characteristics. Children, for example, do not have the rights of adults because their capacities for reason and moral choice are not fully developed; they cannot vote and do not have the freedom of person that their parents do in making choices about where to live, whether to go to school, and so on. Societies strip criminals of basic rights for violating the law, and do so more severely in the case of those regarded as lacking a basic human moral sense. In the United States, they can be deprived even of the right to life for certain kinds of crimes. We do not officially strip Alzheimer's patients of their political rights, but we do restrict their ability to drive and make financial decisions, and in practice they usually cease to exercise their political rights as well.

From a natural-rights perspective, then, one could argue that it is reasonable to assign the unborn different rights from those of either infants or children. A day-old infant may not be capable of reason or moral choice, but it already possesses important elements of the normal human emotional gamut—it can get upset, bond to its mother, expect attention, and the like, in ways that a day-old embryo cannot. It is the violation of the natural and very powerful bonding that takes place between parent and infant, in fact, that makes infanticide such a heinous crime in most societies. That we typically hold funerals after the deaths of infants but not after miscarriages is testimony to the naturalness of this distinction. All of this suggests that it does not make sense to treat embryos as human beings with the same kinds of rights that infants possess.

Against this line of argument, we can pose the following considerations, again not from a religious but from a natural-rights perspective. An embryo may be lacking in some of the basic human characteristics possessed by an infant, but it is also not just another group of cells or tissue, because it has the *potential* to become a full human being. In this respect, it differs from an infant, which also lacks many of the most important characteristics of a normal adult human being, only in the degree to which it has realized its natural potential. This implies that while an embryo can be assigned a lower moral status than an infant, it has a higher moral status than other kinds of cells or tissue that scientists work with. It is therefore reasonable, on nonreligious grounds, to question whether researchers should be free to create, clone, and destroy human embryos at will.

Ontogeny recapitulates phylogeny. We have argued that in the evolutionary process that leads from prehuman ancestor to human beings, there was a qualitative leap that transformed the prehuman precursors of language, reason, and emotion into a human whole that cannot be explained as a simple sum of its parts, and that remains an essentially mysterious process. Something similar happens with the development of every embryo into an infant, child, and adult human being: what starts out as a cluster of organic molecules comes to possess consciousness, reason, the capacity for moral choice, and subjective emotions, in a manner that remains equally mysterious.

Putting these facts together—that an embryo has a moral status somewhere between that of an infant and that of other types of cells and tissue, and that the transformation of the embryo into something with a higher status is a mysterious process—suggests that if we are to do things like harvest stem cells from embryos, we should put a lot of limits and constraints around this activity to make sure that it does not become a precedent for other uses of the unborn that would push the envelope further. To what extent are we willing to create and grow embryos for utilitarian purposes? Supposing some miraculous new cure required cells not from a day-old embryo, but tissue from a month-old fetus? A five-month-old female fetus already has in her ovaries all the eggs she will ever produce as a woman; supposing someone wanted access to them? If we get too used to the idea of cloning embryos for medical purposes, will we know when to stop?

If the question of equality in a future biotech world threatens to tear up the Left, the Right will quite literally fall apart over questions related to human dignity. In the United States, the Right (as represented by the Republican Party) is divided between economic libertarians, who like entrepreneurship and technology with minimal regulation, and social conservatives, many of whom are religious, who care about a range of issues including abortion and the family. The coalition between these two groups is usually strong enough to hold up during elections, but it papers over some fundamental differences in outlook. It is not clear that this alliance will survive the emergence of new technologies that, on the one hand, offer enormous health benefits and money-making opportunities for the biotech industry, but, on the other, require violating deeply held ethical norms.

We are thus brought back to the question of politics and political strategies. For if there is a viable concept of human dignity out there, it needs to be defended, not just in philosophical tracts but in the real world of politics, and protected by viable political institutions. It is to this question that we turn in the final part of this book.

# PART III

·

# WHAT TO DO

# THE POLITICAL CONTROL
# OF BIOTECHNOLOGY

Holy cruelty.—A man who held a newborn child in his hands approached
a holy man. "What shall I do with this child?" he asked; "it is wretched,
misshapen, and does not have life enough to die." "Kill it!" shouted the
holy man with a terrible voice; "and then hold it in your arms for three
days and three nights to create a memory for yourself. Never again will
you beget a child this way when it is not time for you to beget."—When
the man had heard this, he walked away, disappointed, and many people
reproached the holy man because he had counseled cruelty; for he had
counseled the man to kill the child. "But is it not crueler to let it live?"
asked the holy man.

                    Friedrich Nietzsche, *The Gay Science*, Section 731

Some new technologies are frightening from the start and create
an instant consensus on the need to establish political controls
over their development and use. When the first atomic bomb
was detonated at Alamogordo, New Mexico, in the summer of 1945,
not one of the witnesses to the event failed to understand that a terrible new potential for destruction had been created. Nuclear weapons
were thus from the start ringed with political controls: individuals
could not freely develop nuclear technology on their own or traffic in
the parts necessary to create atomic bombs, and, in time, nations that

became signatories to the 1968 nonproliferation treaty agreed to control international trade in nuclear technology.

Other new technologies appear to be much more benign, and consequently subject to little or no regulation. Personal computers and the Internet are examples: these new forms of information technology (IT) promised to create wealth, spread access to information and therefore power around more democratically, and foster community among their users. People had to look hard for downsides to the Information Revolution; what they have found to date are issues like the so-called digital divide (that is, inequality of access to IT) and threats to privacy, neither of which qualify as earth-shaking matters of justice or morality. Despite occasional efforts on the part of the world's more statist societies to try to control the use of IT, it has blossomed in recent years with minimal regulatory oversight on either a national or international level.

Biotechnology falls somewhere between these extremes. Transgenic crops and human genetic engineering make people far more uneasy than do personal computers and the Internet. But biotechnology also promises important benefits for human health and well-being. When presented with an advance like the ability to cure a child of cystic fibrosis or diabetes, people find it difficult to articulate reasons why their unease with the technology should stand in the way of progress. It is easiest to object to a new biotechnology if its development leads to a botched clinical trial or to a deadly allergic reaction to a genetically modified food. But the real threat of biotechnology is far more subtle, and therefore harder to weigh in any utilitarian calculus.

In the face of the challenge from a technology like this, where good and bad are intimately connected, it seems to me that there can be only one possible response: countries must regulate the development and use of technology politically, setting up institutions that will discriminate between those technological advances that promote human flourishing, and those that pose a threat to human dignity and well-being. These regulatory institutions must first be empowered to enforce these discriminations on a national level, and must ultimately extend their reach internationally.

The state of the debate on biotechnology is today polarized between two camps. The first is libertarian, and argues that society should not and cannot put constraints on the development of new technology. This camp includes researchers and scientists who want to push back the frontiers of science, the biotech industry that stands to profit from unfettered technological advance, and, particularly in the United States and Britain, a large group that is ideologically committed to some combination of free markets, deregulation, and minimal government interference in technology.

The other camp is a heterogeneous group with moral concerns about biotechnology, consisting of those who have religious convictions, environmentalists with a belief in the sanctity of nature, opponents of new technology, and people on the Left who are worried about the possible return of eugenics. This group, which ranges from activists like Jeremy Rifkin to the Catholic Church, has proposed banning a wide array of new technologies, from in vitro fertilization and stem cell research to transgenic crops and human cloning.

The debate on biotechnology has to move beyond this polarization. Both approaches—a totally laissez-faire attitude toward biotech development, and the attempt to ban wide swaths of future technology— are misguided and unrealistic. Certain technologies, such as human cloning, deserve to be banned outright, for reasons both intrinsic and tactical. But for most other forms of biotechnology that we see emerging, a more nuanced regulatory approach will be needed. While everyone has been busy staking out ethical positions pro and con various technologies, almost no one has been looking concretely at what kinds of institutions would be needed to allow societies to control the pace and scope of technology development.

It has been a long time since anyone has proposed that what the world needs is more regulation. Regulation—and particularly international regulation—is not something that anyone should call for lightly. Before the Reagan-Thatcher revolutions of the 1980s, many sectors of the economies of North America, Europe, and Japan were vastly overregulated, and many continue to be so today. Regulation brings with it many inefficiencies and even pathologies that are well understood. Research has shown, for example, how government regulators develop a self-interest in promoting their own power and position, even as they make claims to speak in the public interest.[1] Poorly thought out regulation can drive up the costs of doing business enormously, stifle innovation, and lead to the misallocation of resources as businesses seek to avoid burdensome rules. A great deal of innovative work has been done in the past generation on alternatives to formal state regulation—for example, the self-regulation of businesses, and more flexible models for rule generation and enforcement.

The inefficiency of any scheme of regulation is a fact of life. We can try to minimize it by designing institutions that seek to streamline the regulatory process and make it more responsive to changes in technology and social needs, but in the end there are certain types of social problems that can only be addressed through formal government control. Schemes for self-regulation tend to work best in situations in which an industry doesn't produce a lot of social costs (negative externalities, in economists' terminology), in which the issues tend to be technical and apolitical, and in which industry itself has strong incentives to police itself. This is true in international standards setting, coordination of airline traffic routes and payments, product testing, and bank settlements and was at one time true for food safety and medical experimentation.

But it is not true of present-day biotechnology, or of the kinds of biomedical technologies that are likely to appear in the future. While the community of research scientists has in the past done an admirable job in policing itself in such areas as human experimentation and the safety of recombinant DNA technology, there are now too many commercial interests chasing too much money for self-regulation to continue to work well into the future. Most biotechnology companies will simply not have the incentives to observe many of the fine ethical distinctions that need to be made, which means that governments necessarily have to step in to draw up and enforce rules for them.

Many people today believe that biotechnology should not and cannot, as a practical matter, be controlled. Both of these conclusions are wrong, as we will see.

### WHO GETS TO DECIDE?

Who gets to decide whether we will control a new biotechnology, and with what authority?

During the debate in the U.S. Congress on bills to ban human cloning in 2001, Congressman Ted Strickland of Ohio insisted that we be guided strictly by the best available science, and that "We should not allow theology, philosophy, or politics to interfere with the decision we make on this issue."

There are many who would agree with this statement. Opinion polls in most countries show the public holding scientists in much higher regard than politicians, not to mention theologians or philosophers. Legislators, as we well know, like to posture, exaggerate, argue by anecdote, pound the table, and pander. They often speak and act out of ignorance and are at times heavily influenced by lobbyists and entrenched interests. Why should they, rather than the disinterested community of researchers, have final say on highly complex and technical issues like biotechnology? Efforts by politicians to limit what scientists do in their own domain evokes memories of the medieval Catholic Church branding Galileo a heretic for saying the earth revolves around the sun. Since the time of Francis Bacon, the pursuit of scientific research has been seen to carry its own legitimacy as an activity that automatically serves the broader interests of mankind.

This view is, unfortunately, not correct.

Science by itself cannot establish the ends to which it is put. Science can discover vaccines and cures for diseases, but it can also create infectious agents; it can uncover the physics of semiconductors but also the physics of the hydrogen bomb. Science qua science is indifferent to whether data are gathered under rules that scrupulously protect the interests of human research subjects. Data, after all, are data, and better data can often be obtained (as we will see in the section on human experimentation in Chapter 11) by bending the rules or ignoring them altogether. A number of the Nazi doctors who injected concentration camp victims with infectious agents or tortured prisoners by freezing or burning them to death were in fact legitimate scientists who gathered real data that could potentially be put to good use.

It is only "theology, philosophy, or politics" that can establish the ends of science and the technology that science produces, and pronounce on whether those ends are good or bad. Scientists may help establish moral rules concerning their own conduct, but they do so not as scientists but as scientifically informed members of a broader political community. There are very many brilliant, dedicated, energetic, ethical, and thoughtful people within the community of research scientists and doctors working in the field of biomedicine. But their interests do not necessarily correspond to the public interest. Scientists are strongly driven by ambition, and often have pecuniary interests in a particular technology or medicine as well. Hence the question of what we do with biotechnology is a political issue that cannot be decided technocratically.

The answer to the question of who gets to decide on the legitimate and illegitimate uses of science is actually pretty simple, and has been established by several centuries of political theory and practice: it is the democratically constituted political community, acting chiefly through their elected representatives, that is sovereign in these matters and has the authority to control the pace and scope of technological development. While there are all sorts of problems with democratic institutions today, from special-interest lobbying to populist posturing, there is also no obviously better alternative set of institutions that can capture the will of the people in a fair and legitimate way. We can surely hope that politicians make decisions that are informed by a sophisticated understanding of science. History is full of cases where laws were made based on bad science, such as the eugenics legislation passed in the United States and Europe in the early twentieth century. But in the end, science itself is just a tool for achieving human ends; what the political community decides are appropriate ends are not ultimately scientific questions.

When we turn to the question of establishing a regulatory regime for human biotechnology, we face a rather different problem. The issue is not whether it should be scientists or politicians who make choices regarding scientific research, but whether what is best in terms of reproductive decisions should be decided by individual parents or the government. James Watson has argued that it should be individual mothers and not a group of male regulators:

> My principle here is pretty simple: just have most of the decisions made by women as opposed to men. They're the ones who bear children, and men, as you know, often sneak away from children that aren't healthy. We're going to have to feel more responsible for the next generation. I think women should be allowed to make the decisions, and as far as I'm concerned, keep these male doctor committees out of action.[2]

Counterpoising the judgment of male bureaucrats against the concerns of loving mothers is a clever rhetorical strategy, but it is beside the point. Male judges, officers, and social workers (as well as a lot of female ones) already interfere in the lives of women all the time, telling them they can't neglect or abuse their children, that they have to send them to school rather than making them go out to earn money for the family, or that they can't give them drugs or arm them with weapons. The fact that most women will use their authority responsibly doesn't eliminate the need for rules, particularly when technology makes possible all sorts of highly unnatural reproductive possibilities (like cloning) whose ultimate consequences for children may not be healthy.

As outlined in Chapter 6, the automatic community of interest that is assumed to exist between parent and child under natural forms of reproduction may not exist under the new ones. Some have argued that we can presume the consent of a yet-to-be-born child to be free of birth defects or of mental retardation. But can we presume the consent of a child to be a clone, or to be born the biological offspring of two women, or to be born with a nonhuman gene? Cloning in particular raises the prospect that the reproductive decision will suit the interests and convenience of the parent rather than those of the child, and in this case, the state has an obligation to intervene to protect the child.[3]

## CAN TECHNOLOGY BE CONTROLLED?

Even if we decide that technology should legitimately be controlled, we face the problem of whether it can be. Indeed, one of the greatest obstacles to thinking about a regulatory scheme for human biotechnology is the widespread belief that technological advance cannot be regulated, and that all such efforts are self-defeating and doomed to failure.[4] This is asserted gleefully by enthusiasts of particular technologies and by those who hope to profit from them, and pessimistically by those who would like to slow down the spread of potentially harmful technologies. In the latter camp, particularly, there is a kind of defeatism as to the ability of politics to shape the future.

This belief has become particularly strong in recent years because

of the advent of globalization and our recent experience with information technology. No sovereign nation-state, it is said, can regulate or
ban any technological innovation, because the research and development will simply move to another jurisdiction. American efforts to
control data encryption, for example, or French efforts to enforce a
French-language policy on French Web sites, have simply hobbled
technological development in these countries, as developers moved
their operations to more favorable regulatory climates. The only way
to control the spread of technology is to have international agreements on technology-restricting rules, which are extraordinarily difficult to negotiate and even harder to enforce. In the absence of such
international agreements, any nation that chooses to regulate itself
will simply give other nations a leg up.

This kind of pessimism about the inevitability of technological advance is wrong, and it could become a self-fulfilling prophecy if believed by too many people. For it is simply not the case that the speed
and scope of technological development cannot be controlled. There
are many dangerous or ethically controversial technologies that have
been subject to effective political control, including nuclear weapons
and nuclear power, ballistic missiles, biological and chemical warfare
agents, replacement human body parts, neuropharmacological drugs,
and the like, which cannot be freely developed or traded internationally. The international community has regulated human experimentation effectively for many years. More recently, the proliferation of
genetically modified organisms (GMOs) in the food chain has been
stopped dead in its tracks in Europe, with American farmers walking
away from transgenic crops that they had only recently embraced.
One can argue about the rightness of this decision on scientific
grounds, but it proves that the march of biotechnology is not an un-
stoppable juggernaut.

Indeed, the common assumption that it is impossible to control
pornography or political discussion on the Internet is wrong. It is not
possible for a government to shut down every objectionable Web site
around the world, but it is possible to raise the costs of accessing
them for ordinary people who live in their jurisdictions. The Chinese
authorities, for instance, have used their political power effectively to
force Internet companies like Yahoo! and MSN to restrict publication

of unsympathetic stories on their Chinese-language Web sites by simply threatening to revoke their right to operate in China.

Skeptics will argue that none of these efforts to control technology has been successful in the end. Despite the huge diplomatic effort that the West, and especially the United States, has put into nonproliferation, for example, India and Pakistan became the sixth and seventh powers to test nuclear devices openly in the 1990s. While nuclear power for energy generation was slowed down after Three Mile Island and Chernobyl, it is now back on the table because of rising fossil fuel costs and concerns over global warming. Ballistic missile proliferation and the development of weapons of mass destruction continue in places like Iraq and North Korea, while there is a large underground market in drugs, spare body parts, plutonium, and virtually any other illicit commodity one cares to name.

All of this is true enough: no regulatory regime is ever fully leakproof, and if one selects a sufficiently long time frame, most technologies end up being developed eventually. But this misses the point of social regulation: no law is ever fully enforced. Every country makes murder a crime and attaches severe penalties to homicide, and yet murders nonetheless occur. The fact that they do has never been a reason for giving up on the law or on attempts to enforce it.

In the case of nuclear weapons, vigorous nonproliferation efforts on the part of the international community were in fact very successful in slowing down their spread and keeping them out of the hands of countries that might at different points in their histories have been tempted to use them. At the dawn of the nuclear era, in the late 1940s, experts routinely predicted that dozens of countries would possess nuclear weapons in a few years.[5] That only a handful have developed them, and that none of these weapons had been detonated in conflict by the end of the twentieth century, was a remarkable achievement. There are any number of countries that could have developed nuclear weapons but refrained from doing so. Brazil and Argentina, for example, harbored nuclear ambitions when they were both military dictatorships. The nonproliferation regime in which they were enmeshed, however, forced them to keep these programs secret and slowed their development; when both returned to democracy in the 1980s, the programs were shut down.[6]

Nuclear weapons are easier to control than biotechnology for two reasons. First, nuclear weapons development is very expensive and requires large, visible institutions, making their private development very unlikely. Second, the technology is so obviously dangerous that there was a rapid worldwide consensus on the need to control it. Biotechnology, by contrast, can be carried out in smaller, less lavishly funded labs, and there is no similar consensus on its downside risks.

On the other hand, biotechnology does not pose high enforcement hurdles the way nuclear weapons do. A single bomb in the hands of a terrorist group or rogue state like Iraq will pose significant dangers to the world's security. By contrast, an Iraq that can clone Saddam Hussein does not pose much of a threat, unappetizing as that prospect may be. The purpose of a law banning human cloning in the United States would not be undermined if some other countries in the world permitted it, or if Americans could travel abroad to have themselves cloned in such jurisdictions.

The argument that regulation cannot work in a globalized world unless it is international in scope is true enough, but to use this fact to build a case against national-level regulation is to put the cart before the horse. Regulation seldom starts at an international level: nation-states have to develop rules for their own societies before they can even begin to think about creating an international regulatory system.* This is particularly true in the case of a politically, economically, and culturally dominant country like the United States: other countries around the world will pay a great deal of attention to what the United States does in its domestic law. If an international consensus on the regulation of certain biotechnologies is ever to take shape, it is very difficult to see it coming about in the absence of American action at a national level.

In pointing to other cases where technology has been regulated with some success, I do not mean to underestimate the difficulty of creating a similar system for human biotechnology. The international

---

*There are some exceptions to this general rule, such as the case of new or transitional democracies that appeal to international rules on human rights to promote the observance of these rules in their own societies. This analogy is not appropriate in the case of rules for biotechnology, however. International conventions on human rights were established at the instigation of countries that observed these rights and had them codified in their legal systems already.

biotech industry is highly competitive, and companies are constantly searching for the most favorable regulatory climate in which to do their work. Because Germany, with its traumatic history of eugenics, has been more restrictive of genetic research than many developed countries, most German pharmaceutical and biotech companies have moved their labs to Britain, the United States, and other less restrictive countries. In 2000, Britain legalized therapeutic or research cloning and will become a haven for this type of research should the United States join Germany, France, and a number of other countries that do not permit it. Singapore, Israel, and other countries have indicated an interest in pursuing research in stem cells and other niches if the United States continues to restrict its own efforts out of ethical concerns.

The realities of international competition do not mean, however, that the United States or any other country has to fatalistically jump into a technological arms race. We do not know at this point whether an international consensus on the banning or strict regulation of certain technologies, such as cloning and germ-line modification, will emerge, but there is absolutely no reason to rule out the possibility at this early stage in the game.

Take the issue of reproductive cloning—that is, the cloning of a human child. As of this writing (November 2001), twenty-four countries had banned reproductive cloning, including Germany, France, India, Japan, Argentina, Brazil, South Africa, and the United Kingdom. In 1998 the Council of Europe approved an Additional Protocol to its Convention on Human Rights and Dignity with Regard to Biomedicine, banning human reproductive cloning; the document was approved by twenty-four of the council's forty-three member states. The U.S. Congress was just one of a number of other legislatures deliberating on similar measures. The French and German governments have proposed that the United Nations enact a global reproductive cloning ban. Given that Dolly the sheep had been created only four years earlier, it is not surprising that it has taken time for politicians and the law to catch up with technology. But at the moment it appears that much of the world is heading toward a consensus on the illegitimacy of human reproductive cloning. It may be that in a few years, if some crackpot cult like the Raelians wants to clone a child, they will have to travel to North Korea or Iraq to do so.

What are the prospects for the emergence of an international consensus on biotech regulation? It is hard to say at this early point, but it is possible to make some observations about culture and politics with regard to this issue.

There is a continuum of views in the world today concerning the ethicality of certain types of biotechnology and particularly genetic manipulation. At the most restrictive end of this continuum are Germany and other countries in continental Europe that, for historical reasons already mentioned, have been very reluctant to move too far down this road. Continental Europe has also been home to the world's strongest environmental movements, which as a whole have been quite hostile to biotechnology in its various forms.

At the other end of the spectrum are a number of countries in Asia, which for historical and cultural reasons have not been nearly as concerned with the ethical dimension of biotechnology. Much of Asia, for example, lacks religion per se as it is understood in the West—that is, as a system of revealed belief that originates from a transcendental deity. The dominant ethical system in China, Confucianism, lacks any concept of God; folk religions like Taoism and Shinto are animistic and invest both animals and inanimate objects with spiritual qualities; and Buddhism conflates human and natural creation into a single seamless cosmos. Asian traditions such as Buddhism, Taoism, and Shinto tend not to make as sharp an ethical distinction between mankind and the rest of natural creation as does Christianity. That these traditions perceive a continuity between human and nonhuman nature has allowed them to be, as Frans de Waal points out, more sympathetic to nonhuman animals.[7] But it also implies a somewhat lower degree of regard for the sanctity of human life. Consequently, practices such as abortion and infanticide (particularly female infanticide) have been widespread in many parts of Asia. The Chinese government has permitted practices abhorrent in the West, such as the harvesting of organs from executed prisoners, and passed a eugenics law as recently as 1995.

Between continental Europe and Asia on the continuum lie the English-speaking countries, Latin America, and other parts of the world. America and Britain never developed the phobia for genetic research that Germany and France did, and are by virtue of their liberal traditions more skeptical of state regulation. The United States in particular has always been addicted to technological innovation and, for a host of institutional and cultural reasons, is very good at producing it. The American fondness for technology has been strongly reinforced by the information technology revolution of the last two decades, which has convinced many Americans that technology inevitably promises to be individually liberating and personally enriching. Balanced against this is the strength of conservative religious groups in the United States—Protestant, Catholic, and, increasingly, Muslim—that have up to now acted as brakes on uncontrolled technological advance.

Britain has always been closer to America, with its liberal traditions, than to Germany, but it has paradoxically been home to one of the strongest environmental protest movements opposed to GMOs and agricultural biotechnology. There are probably no deep cultural reasons for this; British skepticism about GMOs is more likely traced to the massive regulatory failure represented by mad cow disease, a failure that has left Britain with the largest population of victims to date of the human form of bovine spongiform encephalopathy (BSE), Creutzfeldt-Jacob disease. BSE has nothing to do with biotechnology, of course, but it did reasonably raise doubts in people's minds about the credibility of governments that pronounce on the safety of food products. A generation ago, Americans were much more concerned with threats to the environment and eager to regulate them, based on their recent experiences with Love Canal and other environmental disasters.

If there is any region of the world that is likely to opt out of an emerging consensus on the regulation of biotechnology, it is Asia. A number of Asian countries either are not democracies or lack strong domestic constituencies opposed to certain types of biotechnology on moral grounds. Asian countries like Singapore and South Korea have the research infrastructure to compete in biomedicine, and strong economic incentives to gain market share in biotechnology at the expense of Europe and North America. In the future, biotechnology may become an important fracture line in world politics.

An international consensus on the control of new biomedical technologies will not simply spring into being without a great deal of work on the part of the international community and the leading countries within it. There is no magic bullet for creating such a consensus. It will require the traditional tools of diplomacy: rhetoric, persuasion, negotiation, economic and political leverage. But in this respect the problem is not different from the creation of any other international regime, whether in air traffic, telecommunications, nuclear or ballistic missile proliferation, and the like.

The international governance of human biotechnology does not inevitably mean the creation of a new international organization, expanding the United Nations, or setting up an unaccountable bureaucracy. At the simplest level it can come about through the effort of nation-states to harmonize their regulatory policies. For members of the European Union (EU), this harmonization will presumably already have occurred at a European level.

Take, for example, the international regime governing pharmaceuticals. Every industrialized country has a science-based regulatory agency comparable to the American Food and Drug Administration to oversee the safety and effectiveness of drugs. In Britain it is the Medicines Control Agency, in Japan the Pharmaceutical Affairs Council, in Germany the Bundesinstitut für Arzneimittel und Medizinprodukte, and in France the Agence Française du Médicament. The European Community has sought to harmonize the drug-approval process of its member states since 1965 to avoid the duplication and waste involved in filing multiple applications in different national jurisdictions. This led to the establishment in London in 1995 of the European Medicines Evaluation Agency, which was supposed to provide one-stop drug approval shopping at a European level.[8] At the same time, the European Commission convened a multilateral meeting to broaden harmonization beyond Europe (called the International Conference on Harmonization). Although some Americans have criticized this as an effort by Eurocrats to extend their reach to the United States, it remains a voluntary regime that has received strong support from the pharmaceutical industry because it could lead to substantial increases in efficiency.[9]

Before we can discuss how human biotechnology needs to be regulated in the future, however, we need to understand how it is regulated today, and how the current system came into existence. The picture is extraordinarily complex, particularly when seen on an international level, and is one in which the history of agricultural and human biotechnology have been closely intertwined.

## HOW BIOTECHNOLOGY IS
## REGULATED TODAY

There are many different approaches to regulation, ranging from self-regulation by industry or the scientific community with minimal government oversight, to formal regulation by a statutory agency. Formal regulation can, moreover, be more or less intrusive: at one extreme, there can be a close relationship between regulator and regulatee, which often invites charges of industry "capture," but there can also be highly adversarial relationships, in which the regulating agency imposes detailed (and unwanted) rules on the target industry and is subject to constant litigation. Many of these variants have already been applied to biotechnology.

Take genetic engineering. The development of the underlying technology of recombinant DNA (rDNA), in which genes are spliced (often from one species to another), gave rise to an early and exemplary case of self-regulation by the scientific community. In 1970 Janet Mertz, a researcher at Cold Spring Harbor Laboratory in New York, wanted to splice genes from a monkey virus into a common bacteria, *E. coli*, in order to better understand their function. This led to a dispute between Mertz's supervisor, Paul Berg, and Robert Pollack over the safety of such experiments; Pollack feared they could lead to the creation of a new and highly dangerous microbe.[1]

The eventual result was the Asilomar Conference, held in Pacific Grove, California, in 1975, at which the leading researchers in the field met to devise controls over experiments in the burgeoning field of rDNA.[2] A voluntary ban on this type of research was put into place until the risks could be better appreciated, and a Recombinant DNA Advisory Committee was established by the National Institutes of Health. The NIH published guidelines for NIH-funded research in 1976 that, among other things, required the physical containment of rDNA organisms in the laboratory and restricted their release into the environment.

As it turned out, fears that rDNA research would lead to the creation of dangerous "superbugs" proved unfounded; virtually all the new organisms proved much less robust than their naturally occurring relatives. Based on further research, the NIH began to lift its rules on laboratory containment of new organisms and release into the environment, and thus permitted the emergence of the present-day agricultural biotech industry. In 1983 the NIH approved the first field trial of a genetically modified organism (GMO), the so-called ice-minus microbe, designed to limit frost damage to crops like tomatoes and potatoes. Genetic engineering was controversial from the start; the ice-minus experiment was held up for a number of years in the 1980s because of litigation that charged that the NIH had not complied with the Environmental Protection Agency's decision-making and public-notification guidelines.

## RULES FOR AGRICULTURAL BIOTECHNOLOGY

The current system for regulating agricultural biotechnology in the United States is based on the Coordinated Framework for Regulation of Biotechnology, which was published in 1986 by the White House

Office of Science and Technology Policy. This was the product of a review by a working group set up by the Reagan administration that confronted the issue of whether new laws and institutions were necessary to oversee the emerging biotech industry. The working group decided that GMOs did not represent dramatic new dangers and that the existing regulatory framework was sufficient for dealing with them. Oversight was parceled out to three different agencies on the basis of their existing statutory authority. The Food and Drug Administration (FDA) evaluates the safety of food and food additives; the Environmental Protection Agency checks the consequences of new organisms for the environment; and the Department of Agriculture's Animal and Plant Health Inspection Service oversees the raising or growing of meat and agricultural products.[3]

The American regulatory environment is relatively relaxed and has permitted the field testing and eventual commercialization of a variety of GMOs, including Bt corn, Roundup Ready soybeans, and the so-called Flavr-Savr Tomato.[4] American regulators by and large have not adopted an adversarial relationship with the companies and individuals seeking approval of new GMOs. They do not maintain a strong independent capacity for evaluating the long-term environmental impacts of biotech products but rely instead on assessments provided by the applicants or by outside experts.[5]

The European regulatory environment for biotechnology is considerably more restrictive. This is due in part to political opposition to GMOs, which has been much stronger in Europe than in North America, but also to the fact that all regulation tends to be more cumbersome in Europe because it exists at both national and European levels. There is considerable variation among European Union (EU) member states with regard to the mode and level of biotech regulation. Denmark and Germany have passed relatively stringent national laws regulating safety and ethical aspects of genetic modification; the United Kingdom, by contrast, established a Genetic Manipulation Advisory Group, within the Department of Education and Science, which has maintained a relatively hands-off approach. The French, despite their dirigiste tendencies, relied until 1989 on self-regulation by the French scientific community.[6] By EU rules, individual member states are allowed to be more restrictive than the community as a whole, though the degree to which this is permissible is a matter of dispute. Austria and Luxemburg, for example, have banned the planting of certain genetically modified crops, which is legal in the rest of the EU.[7]

Given the requirement that goods be traded freely within the internal market, the European Commission has been the primary rule-setting body. In 1990 it issued two directives, the first on the contained use of genetically modified microorganisms (Directive 90/219), and the second on the deliberate release into the environment of genetically modified organisms (Directive 90/220).[8] These directives laid the groundwork for evaluating new biotech products on the basis of a "precautionary principle," which asserts in effect that products should be presumed guilty until proven innocent of potentially threatening the environment or public health.[9] These were supplemented in 1997 by Regulation 97/258, which required the labeling of so-called novel food. A further directive on GMOs was adopted by the EU Council of Ministers, requiring strict oversight and labeling of biotech products, tightening up the constraints imposed by earlier legislation. These regulatory requirements have greatly slowed the introduction of GMOs into Europe and have imposed strict labeling requirements on those approved for sale there.

Europeans are not, of course, of one mind on these issues; apart from national differences, there is a substantial divergence of perspective between the powerful European biotech and pharmaceutical industries and groups concerned with the environment and consumer protection. These splits are reflected in the commission itself, with the Directorates-General for Industrial Affairs and Science pushing for looser rules, and the Directorate-General for Environment demanding that environmental concerns be placed above economic interests.[10]

Food safety regulation also exists at an international level. In 1962 the UN Food and Agriculture Organization and the World Health Organization jointly established the Codex Alimentarius Commission, whose mandate was to harmonize existing food safety standards and to develop new international ones. The Codex standards are voluntary, but under the rules of the General Agreement on Tariffs and Trade (GATT) and its successor, the World Trade Organization (WTO), they are used as a reference standard for judging whether a national standard complies with GATT/WTO requirements. The

WTO's Agreement on Sanitary and Phytosanitary Measures (SPS) sets out a number of rules for the establishment of national food safety regulations.[11] If a WTO member imposes food safety standards that are more rigorous than those of the Codex, and they do not seem to be science-based, other members have grounds for challenging them as unfair trade restrictions.

Until the emergence of GMOs, the Codex Alimentarius was regarded as an exemplar of effective international technocratic governance. It gave developing countries with poorly funded regulatory systems a ready-made set of standards and promoted greater global trade in food products. With the rise of biotechnology, however, the Codex's work has become considerably more politicized: critics have charged that its standards are too heavily influenced by the global agricultural and biotech industries, and its work too shielded from public scrutiny.[12]

The environmental dimension of agricultural biotechnology has been addressed at the international level by the Cartagena Protocol on Biosafety, which was adopted by an international conference not in Cartagena (Colombia) but in Montreal, Canada, in January 2000. The protocol allows importing countries to restrict imports of GMOs in the absence of scientific certainty that the product in question will be harmful, and it requires companies wishing to import such products to notify the importing country of the presence of GMOs. The Europeans regarded adoption of the Cartagena Protocol as a victory for the precautionary principle; it will come into force when it is ratified by fifty countries.[13] The United States cannot sign the protocol because it is not a party to the parent Convention on Biological Diversity (the so-called Rio Treaty), though as the largest exporter of GMO products, it may be forced to abide by the protocol's provisions.[14]

The regulatory regime surrounding agricultural biotechnology has been extremely controversial, with the biggest fights occurring between the United States and the EU.[15] The United States has not accepted the precautionary principle as a risk standard, arguing instead that the burden of proof must lie with those who claim that safety or environmental harms exist, rather than with those who claim they do not.[16] The United States has also opposed the mandatory labeling of GMOs, since labeling requirements force an expensive separation of the GMO and non-GMO food processing chains.[17] The United States is particularly concerned that the Cartagena Protocol may undermine the WTO's SPS provisions and provide a legal basis for restrictions on imports of GMO products that are not scientifically based.

There are a number of reasons for this difference in viewpoint. The United States is the world's largest agricultural exporter and was an early adopter of genetically modified crops; it has a lot to lose if importing countries can impose restrictions on GMOs or require costly labeling. American farmers are export-oriented and pro–free trade; European farmers tend to be much more protectionist. There has been little consumer backlash against genetically modified foods in the United States, as there has been in Europe, though some food processors have begun to label GMO products voluntarily. Europe, by contrast, has a much stronger environmental movement, which has been very hostile to biotechnology.

## HUMAN BIOTECHNOLOGY

The regulatory regime for human biotechnology is much less developed than for agricultural biotech, largely because the genetic modification of human beings has not yet arrived as it has for plants and animals. Parts of the existing regulatory structure will be applicable to the new innovations over the horizon; parts are just now being put into place; but the most important elements of a future regulatory system have yet to be invented.

The elements of the existing regulatory structure that are most relevant to human biotechnology developments in the future are the rules concerning the two closely related areas of human experimentation and drug approval.

The evolution of rules concerning human experimentation are interesting not just because they would apply to future experiments with human cloning and germ-line engineering, but also because they represent a case of significant ethical constraints being effectively applied, both nationally and internationally, to scientific research. This case runs counter to the received wisdom concerning regulation: it shows that there is no inevitability to the unfettered advance of science and technology, and it is strongest in the country that is supposedly the most hostile to government regulation, the United States.

Rules regarding human experimentation evolved in tandem with regulation of the drug industry in the United States, and were driven forward in each instance by the revelation of scandal or atrocity. In 1937, 107 deaths resulted from the commercial release of the untested sulfanilamide elixir, which was later found to contain the poison diethylene glycol.[18] This scandal led very quickly to passage of the Food, Drug, and Cosmetic Act of 1938, which still remains the statutory basis for the FDA's regulatory authority over new foods and drugs. The thalidomide scandal of the late 1950s and early 1960s led to passage of the Kefauver Drug Amendments Act of 1962, which tightened the rules governing the "informed consent" of a participant in drug trials. Thalidomide, which had been approved for use in Britain, led to horrifying birth defects in the children of women who had taken it when pregnant. Its approval had been held up by the FDA at the clinical trial stage, but the drug nonetheless led to birth defects among the children of mothers participating in the trials.[19]

Human subjects have been threatened not just by new drugs, but by scientific experimentation more broadly. The United States developed an extensive set of rules protecting human subjects in scientific experiments largely because of the role played by the NIH (and its parent, the U.S. Public Health Service) in funding biomedical research in the postwar period. Again, regulation was driven by scandal and tragedy. The NIH in its early years set up a peer review system for evaluating research proposals but tended to defer to the judgment of the scientific community in establishing acceptable risks to human research subjects. This system proved inadequate with the revelation of the Jewish Chronic Disease Hospital scandal (in which chronically ill and feeble patients were injected with live cancer cells), the Willowbrook scandal (in which mentally retarded children were infected with hepatitis), and the Tuskegee syphilis scandal (in which 400 poor black men diagnosed with syphilis were put under observation but not told of their condition and, in many cases, not treated for it when medications became available).[20] These incidents led in 1974 to new federal regulations protecting human research subjects, and the National Research Act, which created the National Commission for the Protection of Subjects of Biomedical and Behavioral Research.[21] These new laws laid the basis for the current system of Institutional

Review Boards, now required for federally funded research. Even now, the adequacy of these protections has been criticized; the National Bioethics Advisory Commission issued a report in 2001 urging new federal legislation creating a single, strengthened National Office for Human Research Oversight.[22]

Then as now, scientists pursuing ethically questionable research defended their actions on the grounds that the medical benefits that could be derived from their work outweighed possible harms to the research subjects. They also argued that the scientific community alone was best able to judge the risks and benefits of biomedical research, and resisted the intrusion of federal law into their domain.

Rules regarding human experimentation exist on an international level as well. The basic law is the Nuremberg Code, which established the principle that medical experimentation could be performed on a human subject only with the latter's consent.[23] The code grew out of the revelations of the horrifying experiments performed by Nazi doctors on concentration camp inmates during World War II.[24] It had little effect on actual practice in other countries, however, as a recital of later abuses in the United States indicates, and was resisted by many doctors as being too restrictive of valid research.[25]

The Nuremberg Code was largely superseded by the Helsinki Declaration, adopted by the World Medical Association (the global organization representing national medical associations) in 1964. The Helsinki Declaration establishes a number of principles governing experimentation on human subjects, including informed consent, and was better liked by the international medical profession because it was a matter of self-regulation rather than formal international law.[26] Actual practice among developed nations varies a great deal despite these international rules; Japan, for example, has seen a number of cases in the 1990s in which patients were not informed of their conditions or possible treatments by doctors.

Despite variations in practice and occasional lapses, the case of human experimentation shows that the international community is in fact able to place effective limits on the way in which scientific research is conducted, in ways that balance the need for research against respect for the dignity of research subjects. This is an issue that will need to be revisited frequently in the future.

**12**

# POLICIES FOR THE FUTURE

Advances in biotechnology have created gaping holes in the existing regime for the regulation of human biomedicine, which legislatures and administrative agencies around the world have been racing to fill. It is not clear, for example, that the rules for human experimentation described in the last chapter apply to embryos outside the womb. The nature of the players and the flow of money within the biomedical and pharmaceutical communities have also changed, with important implications for any future regulatory system.

One thing is reasonably clear: the time when governments could deal with biotech questions by appointing national commissions that brought scientists together with learned theologians, historians, and bioethicists—groups like the National Bioethics Advisory Commission in the United States and the European Group on Ethics in Science and New Technologies—is rapidly drawing to a close. These commissions played a very useful role in doing the preliminary intellectual spadework of thinking through moral and social implications of biomedical research. But it is time to move from thinking to acting, from recommending to legislating. We need institutions with real enforcement powers.

The community of bioethicists that has grown up in tandem with the biotech industry is in many respects a double-edged sword. On the one hand, it has played an extremely useful function by raising doubts and questions about the wisdom and morality of certain technological innovations. On the other hand, many bioethicists have become nothing more than sophisticated (and sophistic) justifiers of whatever it is the scientific community wants to do, having enough knowledge of Catholic theology or Kantian metaphysics to beat back criticisms by anyone coming out of these traditions who might object more strenuously. The Human Genome Project from the beginning devoted 3 percent of its budget to studying the Ethical, Social, and Legal Implications of genetic research. This can be regarded as commendable concern for the ethical dimensions of scientific research, or else as a kind of protection money the scientists have to pay to keep the true ethicists off their backs. In any discussion of cloning, stem cell research, germ-line engineering, and the like, it is usually the professional bioethicist who can be relied on to take the most permissive position of anyone in the room.* But if the ethicist isn't going to tell you that you can't do something, who will?

---

*This phenomenon is a common one and is known as regulatory "capture," whereby the group that is supposed to be overseeing the activities of an industry becomes an agent for the industry. This happens for many reasons, including the dependence of the regulators on the regulatees for money and information. In addition, there are the career incentives that most professional bioethicists face. Scientists do not usually have to worry about winning the respect of ethicists, particularly if they are Nobel Prize winners in molecular biology or physiology. On the other hand, ethicists face an uphill struggle winning the respect of the scientists they must deal with, and are hardly likely to do so if they tell them they are morally wrong or if they depart significantly from the materialist worldview that the scientists hold dear.

A number of countries have in fact moved beyond the stage of national commissions and study groups to actual legislation. One of the first and most contentious policy issues legislators have tried to grapple with concerns the uses that may be made of human embryos. This issue touches on a whole host of medical practices and procedures, both existing today and yet to be developed. These include abortion, in vitro fertilization, preimplantation diagnosis and screening, sex selection, stem cell research, cloning for reproductive and research purposes, and germ-line engineering. There are a huge number of permutations and combinations of possible rules that societies can establish regarding embryos. For example, one can imagine permitting them to be aborted or discarded by in vitro fertilization clinics, yet not created deliberately for research purposes nor selected for sex or other characteristics. Formulation and enforcement of these rules will constitute the substance of any future regulatory system for human biotechnology. There are at present a wide variety of national-level rules regarding human embryos. To date (November 2001), sixteen countries have passed laws regulating human embryo research, including France, Germany, Austria, Switzerland, Norway, Ireland, Poland, Brazil, and Peru (despite the fact that in France abortion is legal). In addition, Hungary, Costa Rica, and Ecuador implicitly restrict research by conferring on embryos a right to life. Finland, Sweden, and Spain permit embryo research, but only on extra embryos left over from in vitro fertilization clinics. Germany's laws are among the most restrictive; since passage of the 1990 Act for the Protection of Embryos (Gesetz zum Schutz von Embryonen), a number of areas have been regulated, including abuse of human embryos, sex selection, artificial modification of human germ-line cells, cloning, and the creation of chimeras and hybrids.

Britain in 1990 passed the Fertilisation and Embryology Act, which established one of the most clear-cut legal frameworks in the world for the regulation of embryo research and cloning. This act was thought to ban reproductive cloning while permitting research cloning, though in 2001 a British court ruled reproductive cloning would actually be permitted under a loophole that the government moved quickly to try to close.[1] Given the lack of consensus across the continent on this issue, there has been no action on the European level to regulate embryo research apart from the creation of the European Group on Ethics in Science and New Technologies.[2]

Embryo research is only the beginning of a series of new developments created by technology for which societies have to decide on rules and regulatory institutions. Others that will come up sooner or later include:

- *Preimplantation diagnosis and screening.* This group of technologies, in which multiple embryos are screened genetically for birth defects and other characteristics, is the beginning point of "designer babies" and will arrive much sooner than human germ-line engineering. Indeed, such screening has already been performed for children of parents susceptible to certain genetic diseases. In the future, do we want to permit parents to screen and selectively implant embryos on the basis of sex; intelligence; looks; hair, eye, or skin color; sexual orientation; and other characteristics once they can be identified genetically?

- *Germ-line engineering.* If and when human germ-line engineering arrives, it will raise the same issues as preimplantation diagnosis and screening, but in a more extreme form. Preimplantation diagnosis and screening is limited by the fact that there will always be a limited number of embryos from which to choose, based on the genes of the two parents. Germ-line engineering will expand the possibilities to include virtually any other genetically governed trait, provided it can be identified successfully, including traits that come from other species.

- *The creation of chimeras using human genes.* Geoffrey Bourne, former director of the Emory University primate center, once stated that "it would be very important scientifically to try to produce an ape-human cross." Other researchers have suggested using women as "hosts" for the embryos of chimpanzees or gorillas.[3] One biotech company, Advanced Cell Technology, reported that it had successfully transferred human DNA into a cow's egg and gotten it to grow into a blastocyst before it was destroyed. Scientists have been deterred from doing research in this area for fear of bad publicity, but in the United States such work is not illegal. Will we permit the creation of hybrid creatures using human genes?

- *New psychotropic drugs.* In the United States, the Food and Drug Administration (FDA) regulates therapeutic drugs, while the Drug Enforcement Administration (DEA) and the states regulate illegal narcotics such as heroin, cocaine, and marijuana. Societies will have to make decisions on the legality and extent of permissible use of future generations of neuropharmacological agents. In the case of prospective drugs that improve memory or other cognitive skills, they will have to decide on the desirability of enhancement uses and how they are to be regulated.

## WHERE DO WE DRAW RED LINES?

Regulation is essentially the act of drawing a series of red lines that separate legal from proscribed activities, based on a statute that defines the area in which regulators can exercise some degree of judgment. With the exception of some die-hard libertarians, most people reading the above list of innovations that may be made possible by biotechnology will probably want to see some red lines drawn.

There are certain things that should be banned outright. One of them is reproductive cloning—that is, cloning with the intent of producing a child.[4] The reasons for this are both moral and practical, and go way beyond the National Bioethics Advisory Commission's concerns that human cloning cannot now be done safely.

The moral reasons have to do with the fact that cloning is a highly unnatural form of reproduction that will establish equally unnatural relationships between parents and children.[5] A cloned child will have a very asymmetrical relationship with his or her parents. He or she will be both child and twin of the parent from whom his or her genes come, but will not be related to the other parent in any way. The unrelated parent will be expected to nurture a younger version of his or her spouse. How will that parent look upon the clone when he or she reaches sexual maturity? Nature, for all of the reasons explicated in earlier chapters of this book, is a valid point of reference for our

values and should not be discarded as a standard for parent-child relationships lightly. While it is possible to come up with some sympathetic scenarios in which cloning might be justified (for example, a
Holocaust surviver with no other way of continuing the family line),
they do not constitute a sufficiently strong societal interest to justify a
practice that on the whole would be harmful.[6]

Beyond these considerations inherent to cloning itself, there are a
number of practical concerns. Cloning is the opening wedge for a series of new technologies that will ultimately lead to designer babies
and one that is likely to become feasible much sooner than genetic
engineering. If we get used to cloning in the near term, it will be
much harder to oppose germ-line engineering for enhancement purposes in the future. It is important to lay down a political marker at an
early point to demonstrate that the development of these technologies
is not inevitable, and that societies can take some measure of control
over the pace and scope of technological advance. There is no strong
constituency in favor of cloning in any country. It is also an area
where considerable international consensus exists in opposition to
the procedure. Cloning therefore represents an important strategic
opportunity to establish the possibility of political control over biotechnology.

But while a broad-brush ban is appropriate in this case, it will not
be a good model for the control of future technologies. Preimplantation diagnosis and screening have begun to be used today to ensure
the birth of children free of genetic diseases. The same technology
can be used for less laudable purposes, such as sex selection. What
we need to do in this case is not ban the procedure but regulate it,
drawing red lines not around the procedure itself but within its range
of possible uses to distinguish between what is legitimate and what is
illegitimate.

One obvious way to draw red lines is to distinguish between therapy and enhancement, directing research toward the former while
putting restrictions on the latter. The original purpose of medicine is,
after all, to heal the sick, not to turn healthy people into gods. We
don't want star athletes to be hobbled by bad knees or torn ligaments,
but we also don't want them to compete on the basis of who has taken
the most steroids. This general principle would allow us to use

biotechnologies to, for example, cure genetic diseases like Huntington's chorea or cystic fibrosis, but not to make our children more intelligent or taller.

The distinction between therapy and enhancement has been attacked on the grounds that there is no way to distinguish between the two in theory, and therefore no way of discriminating in practice. There is a long tradition, argued most forcefully in recent years by the French postmodernist thinker Michel Foucault,[7] which maintains that what society considers to be pathology or disease is actually a socially constructed phenomenon in which deviation from some presumed norm is stigmatized. Homosexuality, to take one example, was long considered unnatural and was classified as a psychiatric disorder until the latter part of the twentieth century, when it was depathologized as part of the growing acceptance of gayness in developed societies. Something similar can be said of dwarfism: human heights are distributed normally, and it is not clear at what point in the distribution one becomes a dwarf. If it is legitimate to give growth hormone to a child who is in the bottom 0.5 percentile for height, who's to say that you can't also prescribe it for someone who is in the fifth percentile, or for that matter in the fiftieth?[8] Geneticist Lee Silver makes a similar argument about future genetic engineering, saying that it is impossible to draw a line between therapy and enhancement in an objective manner: "in every case, genetic engineering will be used to add something to a child's genome that didn't exist in the genomes of either of its parents."[9]

While it is the case that certain conditions do not lend themselves to neat distinctions between pathological and normal, it is also true that there is such a thing as health. As Leon Kass has argued, there is a natural functioning to the whole organism that has been determined by the requirements of the species' evolutionary history, one that is not simply an arbitrary social construction.[10] It has often seemed to me that the only people who can argue that there is no difference in principle between disease and health are those who have never been sick: if you have a virus or fracture your leg, you know perfectly well that something is wrong.

And even in the cases where the borderline between sickness and health, therapy and enhancement, is murkier, regulatory agencies are routinely able to make these distinctions in practice. Take the case of Ritalin. As noted in Chapter 3, the underlying "disease" that Ritalin is supposed to treat, attention deficit–hyperactivity disorder (ADHD), is most likely not a disease at all but simply the label that we put on people who are in the tail of a normal distribution of behavior related to focus and attention. This is in fact a classic case of the social construction of pathology: ADHD was not even in the medical lexicon a couple of generations ago. There is, correspondingly, no neat line between what one might label the therapeutic and enhancement uses of Ritalin. At one end of the distribution, there are children almost anyone would say are so hyperactive that normal functioning is impossible for them, and it is hard to object to treating them with Ritalin. At the other end of the distribution are children who have no trouble whatsoever concentrating or interacting, for whom taking Ritalin might be an enjoyable experience that would give them a high just like any other amphetamine. But they would be taking the drug for enhancement rather than for therapeutic reasons, and thus most people would want to prevent them from doing so. What makes Ritalin controversial is all the children in the middle, who meet some but not all of the diagnostic criteria specified in the *Diagnostic and Statistical Manual of Mental Disorders* for the disease and who nonetheless are prescribed the drug by their family physician.

If there was ever a case, in other words, where the distinction between pathology and health in diagnosis, and therapy and enhancement in treatment, is ambiguous, it is ADHD and Ritalin. And yet, regulatory agencies *make and enforce this distinction all the time.* The DEA classifies Ritalin as a Schedule II pharmaceutical that can only be taken for therapeutic purposes with a doctor's prescription; it clamps down on Ritalin's recreational (that is to say, enhancement) use as an amphetamine. That the boundary between therapy and enhancement is unclear does not make the distinction meaningless. My own strong feeling is that the drug is overprescribed in the United States and used in situations in which parents and teachers ought to employ more traditional means of engaging children and shaping their characters. But the current regulatory system, for all its faults, is better than a situation in which Ritalin is either banned altogether or else sold over the counter like cough medicine.

Regulators are called on all the time to make complex judgments that cannot be held up to precise theoretical scrutiny. What constitutes a "safe" level of heavy metals in the soil, or sulfur dioxide in the atmosphere? How does a regulator justify pushing down the level of a particular toxin in drinking water from fifty to five parts per million, when he or she is trading off health consequences against compliance costs? These decisions are always controversial, but in a sense they are easier to make in practice than in theory. For in practice, a properly functioning democratic political system allows people with a stake in the regulator's decision to push and shove against one another until a compromise is reached.

Once we agree in principle that we will need a capability to draw red lines, it will not be a fruitful exercise to spend a lot of time arguing precisely where they should be placed. As in other areas of regulation, many of these decisions will have to be made on a trial-and-error basis by administrative agencies, based on knowledge and experience not available to us at present. What is more important is to think about the design of institutions that can make and enforce regulations on, for example, the use of preimplantation diagnosis and screening for therapeutic rather than enhancement purposes, and how those institutions can be extended internationally.

As noted at the beginning of this chapter, action has to begin with legislatures stepping up to the plate and establishing rules and institutions. This is easier said than done: biotechnology is a technically complex and demanding subject, one moreover that is changing every day, with a wide variety of interest groups pulling in different directions. The politics of biotechnology does not fall into familiar political categories; if one is a conservative Republican or a left-wing Social Democrat, it is not immediately obvious how one should vote on a bill to permit so-called therapeutic cloning or stem cell research. For these reasons many legislators would rather duck the issue, hoping it will get resolved in some other way.

But to not act under conditions of rapid technological change is in effect to make a decision legitimizing that change. If legislators in democratic societies do not face up to their responsibilities, other institutions and actors will make the decisions for them.

This is particularly true given the peculiarities of the American political system. In the past, it has been the case that the courts have stepped into controversial areas of social policy when the legislature failed to act to negotiate acceptable political rules. In the absence of congressional action on an issue like cloning, it is conceivable that the courts at some later point may be tempted or compelled to step into the breach and discover, for example, that human cloning or research on cloning is a constitutionally protected right. This was a very poor approach to the formulation of law and public policy in the past, one that tainted policies, such as the legalization of abortion, that more properly should have been enacted legislatively. On the other hand, if the American people clearly express their will on human cloning through their democratically elected representatives, the courts will be reluctant to thwart their will through discovery of a new right.

If the legislature does act to put further regulatory controls on human biotechnology, it will face large questions concerning the design of the requisite institutions to implement them. The same issue came up for the United States and the European Community in the 1980s when agricultural biotechnology appeared on the scene: Do we use existing regulatory bodies to do the job, or are the new technologies sufficiently different so that entirely new agencies are required? In the American case, the Reagan administration decided that agricultural biotech did not represent a sufficiently radical break with the past to merit regulation on the basis of process rather than product. It therefore decided to leave regulatory authority with existing agencies like the FDA and the Environmental Protection Agency (EPA), on the basis of their statutory authority. The Europeans, by contrast, decided to regulate on the basis of process and therefore had to create new regulatory procedures for handling biotech products.

All countries face similar decisions today concerning human biotechnology. In the United States, it would be possible to leave regulatory authority with existing institutions like the FDA, the NIH, or consultative groups like the Recombinant DNA Advisory Committee (RAC). It is prudent to be conservative in the creation of new regulatory institutions and additional layers of bureaucracy. On the other hand, there are a number of reasons for thinking that we need to establish new institutions to deal with the challenges of the coming biotech revolution. Not to do so would be like trying to use the Interstate Commerce Commission, which was responsible for regulating trucks, to oversee civil aviation when that industry came into being, rather than creating a separate Federal Aviation Administration.

Let us consider first the case of the United States. An initial reason that existing U.S. institutions are probably not sufficient to regulate future human biotechnology is the question of their narrow mandate. Human biotechnology differs substantially from agricultural biotechnology insofar as it raises a host of ethical questions related to human dignity and human rights that are not an issue for GMOs. While people object to genetically engineered crops on ethical grounds, the most vociferous complaints have had to do with their possible negative consequences for human health and their environmental impact. This is precisely what existing regulatory institutions like the FDA, the EPA, and the U.S. Department of Agriculture have been set up to do. They can be criticized for having the wrong standards or for not being sufficiently cautious, but they are not operating outside their regulatory mandate when they take on genetically modified foods.

Let us suppose that Congress legislatively distinguishes between therapeutic and enhancement uses of preimplantation diagnosis and screening. The FDA is not set up to make politically sensitive decisions concerning the point at which selection for characteristics like intelligence and height ceases to be therapeutic and becomes enhancing, or whether these characteristics can be considered therapeutic at all. The FDA can disapprove a procedure only on the grounds of effectiveness and safety, but there will be many safe and effective procedures that will nonetheless require regulatory scrutiny. The limits of the FDA's mandate are already evident: it has asserted a right to regulate human cloning on the legally questionable grounds that a cloned child constitutes a medical "product" over which it has authority.

One can always try to amend and expand the FDA's charter, but past experience shows that it is very difficult to change the organizational culture of agencies with a long history.[11] Not only will the agency resist taking on new duties, but a shifting mandate will likely mean it will do its old job less well. This implies the need to create a new agency to oversee the approval of new medicines, procedures,

and technologies for human health. In addition to having a broader
mandate, this new authority would have to have different staffing. It
would have to include not just the doctors and scientists who staff
the FDA and oversee clinical trials for new drugs, but other societal
voices that are prepared to make judgments about the technology's social and ethical implications.

A second reason that existing institutions are probably not sufficient to regulate biotechnology in the future has to do with changes
that have taken place in the research community and the biotech/
pharmaceutical industry as a whole over the past generation. There
was a period up through the early 1990s when virtually all biomedical
research in the United States was funded by the NIH or another federal government agency. This meant that the NIH could regulate that
research through its own internal rule-making authority, as in the case
of rules concerning human experimentation. Government regulators
could work hand in glove with committees of scientific insiders, like
the RAC, and be reasonably sure that no one in the United States
was doing dangerous or ethically questionable research.

None of this holds true any longer. While the federal government remains the largest source of research funding, there is a huge
amount of private investment money available to sponsor work in new
biotechnologies. The U.S. biotech industry by itself spent nearly
$11 billion on research in 2000, employs over 150,000 people, and has
doubled in size since 1993. Indeed, the massive government-funded
Human Genome Project was upstaged by Craig Venter's privately
held Celera Genomics in the race to map the human genome. The
first embryonic stem cell lines were cultivated by James Thompson at
the University of Wisconsin, using nongovernment funding in order
to comply with the ban on federally funded research that would harm
embryos. Many of the participants at a workshop held on the twenty-
fifth anniversary of the Asilomar Conference on rDNA concluded
that while the RAC had served an important function in its day, it
could no longer monitor or police the present-day biotech industry. It
has no formal enforcement powers and can bring to bear only the
weight of opinion within the elite scientific community. The nature of
that community has changed over time as well: there are today many
fewer "pure" researchers, with no ties to the biotech industry or commercial interests in certain technologies.[12]

This means that any new regulatory agency not only would have to have a mandate to regulate biotechnology on grounds broader than efficacy and safety but also would have to have statutory authority over all research and development, and not just research that is federally funded. Such an agency, the Human Fertilisation and Embyology Authority, has already been created in Britain for this purpose. Unification of regulatory powers into a single new agency will end the practice of complying with federal funding restrictions by finding private sponsors and, it is hoped, will shed a more uniform light on the whole biotech sector.

What are the prospects for the United States and other countries putting into place a regulatory system of the kind just outlined?[13] There will be formidable political obstacles to creating new institutions. The biotech industry is strongly opposed to regulation (if anything, it would like to see FDA rules loosened), as is, by and large, the community of research scientists. Most would prefer regulation to take place within their own communities, outside the scope of formal law. They are joined in this by advocacy groups representing patients, the elderly, and others with an interest in promoting cures for various diseases, and together these groups form a very powerful political coalition.

There are reasons the biotech industry should consider actively promoting the right kind of formal regulation of human biotechnology, however, out of simple long-term self-interest. For that, it needs to look no further than what happened to agricultural biotechnology, which is a good object lesson in the political pitfalls of advancing a new technology too quickly.

At the beginning of the 1990s, Monsanto, a leading innovator in agricultural biotechnology, considered asking the first Bush administration for stronger formal regulatory rules for its genetically engineered products, including labeling requirements. A change of leadership at the top scuttled this initiative, however, on the grounds that there was no scientific evidence of health risks, and the firm introduced a series of new GMOs that were quickly adopted by American farmers. What the company failed to anticipate was the political backlash that arose in Europe against GMOs, and the strict labeling requirements that the European Union imposed in 1997 for genetically modified foods imported into Europe.[14]

Monsanto and other American firms railed at the Europeans for being unscientific and protectionist, but Europe had sufficient market power to enforce its rules on American exporters. American farmers, without a means of separating genetically modified from non–genetically modified foods, found themselves closed out of important export markets. They responded by planting fewer genetically modified crops after 1997 and charging that they had been misled by the biotech industry. In retrospect, Monsanto executives realized that they had made a big mistake by not working earlier to establish an acceptable regulatory environment that would assure consumers of the safety of their products, even if this did not appear to be scientifically necessary.

The history of pharmaceutical regulation was driven by horror stories like the sulfanilamide elixir and thalidomide. It may be the case that regulations concerning human cloning will have to await the birth of a horribly deformed child who is the product of an unsuccessful cloning attempt. The biotech industry needs to consider whether it is better to anticipate such problems now and work toward formulating a system that serves its interests by assuring people of the safety and ethical nature of its products, or wait until there is a huge public outcry following an outrageous accident or horrifying experiment.

## THE BEGINNING OF POSTHUMAN HISTORY?

The American regime was built, beginning in 1776, on a foundation of natural right. Constitutional government and a rule of law, by limiting the arbitrary authority of tyrants, would protect the kind of freedom that human beings by nature enjoyed. As Abraham Lincoln pointed out eighty-seven years later, it was also a regime dedicated to the proposition that all men are created equal. There would be an equality of freedom only because there was a natural equality of man; or, to put it more positively, the fact of natural equality demanded an equality of political rights.

Critics have pointed out that the United States has never lived up to this ideal of an equality of freedom and has, in its history, excluded entire groups from it. Defenders of the American regime have, more correctly in my view, pointed out that the principle of equal rights has driven a steady expansion in the circle of those entitled to rights. Once it was established that all human beings have natural rights, the big arguments in American political history have been over who falls within that charmed circles of "men" who were said by the Declaration to be created equal. The circle did not initially encompass women, or blacks, or white men without property; however, it was slowly but surely expanded to encompass them in time.

Whether the participants in these arguments recognized it or not, they all had at least an implicit idea of what the "essence" of a human being was and therefore a ground for judging whether one or another individual qualified. Human beings on the surface look, speak, and act very differently from one another, so much of this argument revolved around the question of whether those apparent differences were ones of convention only, or whether they were rooted in nature.

Modern natural science has cooperated to some extent in expanding our view of who qualifies as a human being because it has tended to show that most of the apparent differences between human beings are conventional rather than natural. Where there *are* natural differences, as between men and women, they have been shown to affect nonessential qualities that do not have a bearing on political rights.

So, despite the poor repute in which concepts like natural rights are held by academic philosophers, much of our political world rests on the existence of a stable human "essence" with which we are endowed by nature, or rather, on the fact that we believe such an essence exists.

We may be about to enter into a posthuman future, in which technology will give us the capacity gradually to alter that essence over time. Many embrace this power, under the banner of human freedom. They want to maximize the freedom of parents to choose the kind of children they have, the freedom of scientists to pursue research, and the freedom of entrepreneurs to make use of technology to create wealth.

But this kind of freedom will be different from all other freedoms that people have previously enjoyed. Political freedom has heretofore meant the freedom to pursue those ends that our natures had established for us. Those ends are not rigidly determined; human nature is very plastic, and we have an enormous range of choices conformable with that nature. But it is not infinitely malleable, and the elements that remain constant—particularly our species-typical gamut of emotional responses—constitute a safe harbor that allows us to connect, potentially, with all other human beings.

It may be that we are somehow destined to take up this new kind of freedom, or that the next stage of evolution is one in which, as some have suggested, we will deliberately take charge of our own biological makeup rather than leaving it to the blind forces of natural selection. But if we do, we should do it with eyes open. Many assume that the posthuman world will look pretty much like our own—free, equal, prosperous, caring, compassionate—only with better health care, longer lives, and perhaps more intelligence than today.

But the posthuman world could be one that is far more hierarchical and competitive than the one that currently exists, and full of social conflict as a result. It could be one in which any notion of "shared humanity" is lost, because we have mixed human genes with those of so many other species that we no longer have a clear idea of what a human being is. It could be one in which the median person is living well into his or her second century, sitting in a nursing home hoping for an unattainable death. Or it could be the kind of soft tyranny envisioned in *Brave New World*, in which everyone is healthy and happy but has forgotten the meaning of hope, fear, or struggle.

We do not have to accept any of these future worlds under a false banner of liberty, be it that of unlimited reproductive rights or of unfettered scientific inquiry. We do not have to regard ourselves as slaves to inevitable technological progress when that progress does not serve human ends. True freedom means the freedom of political communities to protect the values they hold most dear, and it is that freedom that we need to exercise with regard to the biotechnology revolution today.

# NOTES

## EPIGRAPH

1. The context for this quotation is the following: "From now on there will be more favorable preconditions for more comprehensive forms of dominion, whose like has never yet existed. And even this is not the most important thing; the possibility has been established for the production of international racial unions whose task will be to rear a master race, the future 'masters of the earth';—a new, tremendous aristocracy, based on the severest self-legislation, in which the will of philosophical men of power and artists-tyrants will be made to endure for millennia—a higher kind of man who, thanks to their superiority in will, knowledge, riches, and influence, employ democratic Europe as their most pliant and supple instrument for getting hold of the destinies of the Earth, so as to work as artists upon 'man' himself."

## CHAPTER 1: A TALE OF TWO DYSTOPIAS

1. Martin Heidegger, *Basic Writings* (New York: Harper and Row, 1957), p. 308.

2. Peter Huber, *Orwell's Revenge: The 1984 Palimpsest* (New York: Free Press, 1994), pp. 222–228.

3. Leon Kass, *Toward a More Natural Science: Biology and Human Affairs* (New York: Free Press, 1985), p. 35.

4. Bill Joy, "Why the Future Doesn't Need Us," *Wired* 8 (2000): 238–246.

5. Tom Wolfe, "Sorry, but Your Soul Just Died," *Forbes ASAP*, December 2, 1996.

6. Letter to Roger C. Weightman, June 24, 1826, in *The Life and Selected Writings of Thomas Jefferson*, Thomas Jefferson (New York: Modern Library, 1944), pp. 729–730.

7. Francis Fukuyama, *The End of History and the Last Man* (New York: Free Press, 1992).

8. Ithiel de Sola Pool, *Technologies of Freedom* (Cambridge, Mass.: Harvard/Belknap, 1983).

9. On this point, see Leon Kass, "Introduction: The Problem of Technology," in *Technology in the Western Political Tradition*, ed. Arthur M. Melzer et al. (Ithaca, N.Y.: Cornell University Press, 1993), pp. 10–14.

10. See Francis Fukuyama, "Second Thoughts: The Last Man in a Bottle," *The National Interest*, no. 56 (Summer 1999): 16–33.

## CHAPTER 2: SCIENCES OF THE BRAIN

1. Quote taken from the e-biomed home page, http://www.liebertpub.com/ebi/default1.asp.

2. For the application of genomics to the study of the mind, see Anne Farmer and Michael J. Owen, "Genomics: The Next Psychiatric Revolution?," *British Journal of Psychiatry* 169 (1996): 135–138. See also Robin Fears, Derek Roberts, et al., "Rational or Rationed Medicine? The Promise of Genetics for Improved Clinical Practice," *British Medical Journal* 320 (2000): 933–995; and C. Thomas Caskey, "DNA-Based Medicine: Prevention and Therapy," in Daniel J. Kevles and Leroy Hood, eds., *The Code of Codes: Scientific and Social Issues in the Human Genome Project* (Cambridge, Mass.: Harvard University Press, 1992).

3. For an overview of this debate, see Frans de Waal, "The End of Nature versus Nurture," *Scientific American* 281 (1999): 56–61.

4. Madison Grant, *The Passing of the Great Race; or, the Racial Basis of European History*, 4th ed., rev. (New York: Charles Scribner's Sons, 1921).

5. Jay K. Varma, "Eugenics and Immigration Restriction: Lessons for Tomorrow," *Journal of the American Medical Association* 275 (1996): 734.

6. See, for example, Ruth Hubbard, "Constructs of Genetic Difference: Race and Sex," in Robert F. Weir and Susan C. Lawrence, eds., *Genes, Humans, and Self-Knowledge* (Iowa City: University of Iowa Press, 1994), pp. 195–205; and Ruth Hubbard, *The Politics of Women's Biology* (New Brunswick, N.J.: Rutgers University Press, 1990).

7. Carl C. Brigham, *A Study of American Intelligence* (Princeton, N.J.: Princeton University Press, 1923).

8. For an argument for the continuity between biology and culture, see Edward O. Wilson, *Consilience: The Unity of Knowledge* (New York: Knopf, 1998), pp. 125–130.

9. Margaret Mead, *Coming of Age in Samoa: A Psychological Study of Primitive Youth for Western Civilization* (New York: William Morrow, 1928).

10. Donald Brown, *Human Universals* (Philadelphia: Temple University Press, 1991), p. 10.

11. Nicholas Wade, "Of Smart Mice and Even Smarter Men," *The New York Times*, September 7, 1999, p. F1.

12. Matt Ridley, *Genome: The Autobiography of a Species in 23 Chapters* (New York: HarperCollins, 2000), p. 137.

13. Luigi Luca Cavalli-Sforza, *Genes, Peoples, and Languages* (New York: North Point Press, 2000), and, with Francesco Cavalli-Sforza, *The Great Human Diasporas: The History of Diversity and Evolution* (Reading, Mass.: Addison-Wesley, 1995).

14. Genetic factors have also been said to play a role in alcoholism. See C. Cloninger, M. Bohman, et al., "Inheritance of Alcohol Abuse: Crossfostering Analysis of Alcoholic Men," *Archives of General Psychiatry* 38 (1981): 861–868.

15. Charles Murray and Richard J. Herrnstein, *The Bell Curve: Intelligence and Class Structure in American Life* (New York: Free Press, 1995).

16. Charles Murray, "IQ and Economic Success," *Public Interest* 128 (1997): 21–35.

17. Arthur R. Jensen, "How Much Can We Boost IQ and Scholastic Achievement?," *Harvard Educational Review* 39 (1969): 1–123.

18. See, passim, Claude S. Fischer et al., *Inequality by Design: Cracking the Bell Curve Myth* (Princeton, N.J.: Princeton University Press, 1996).

19. Robert G. Newby and Diane E. Newby, "The Bell Curve: Another Chapter in the Continuing Political Economy of Racism," *American Behavioral Scientist* 39 (1995): 12–25.

20. Stephen J. Rosenthal, "The Pioneer Fund: Financier of Fascist Research," *American Behavioral Scientist* 39 (1995): 44–62.

21. On testing more broadly, see Nicholas Lemann, *The Big Test: The Secret History of the American Meritocracy* (New York: Farrar, Straus and Giroux, 1999).

22. Francis Galton, *Hereditary Genius: An Inquiry into Its Laws and Consequences* (New York: Appleton, 1869).

23. Karl Pearson, *National Life from the Standpoint of Science*, 2d ed. (Cambridge: Cambridge University Press, 1919), p. 21.

24. Stephen Jay Gould, *The Mismeasure of Man* (New York: W. W. Norton, 1981).

25. Leon Kamin, *The Science and Politics of IQ* (Potomac, Md.: L. Erlbaum Associates, 1974).

26. Richard C. Lewontin, Steven Rose, et al., *Not in Our Genes: Biology, Ideology, and Human Nature* (New York: Pantheon Books, 1984). For a discussion of this debate, see Thomas J. Bouchard, Jr., "IQ Similarity in Twins Reared Apart: Findings and Responses to Critics," in Robert J. Sternberg and Elena L. Grigorenko, eds., *Intelligence, Heredity, and Environment* (Cambridge: Cambridge University Press, 1997); and Thomas J. Bouchard, Jr., David T. Kykken, et al., "Sources of Human Psychological Differences: The Minnesota Study of Twins Reared Apart," *Science* 226 (1990): 223–250.

27. Robert B. Joynson, *The Burt Affair* (London: Routledge, 1989); and R. Fletcher, "Intelligence, Equality, Character, and Education," *Intelligence* 15 (1991): 139–149.

28. Robert Plomin, "Genetics and General Cognitive Ability," *Nature* 402 (1999): C25–C44.

29. See, inter alia, Howard Gardner, *Frames of Mind: The Theory of Multiple Intelligences* (New York: Basic Books, 1983); and *Multiple Intelligences: The Theory in Practice* (New York: Basic Books, 1993).

30. See Bernie Devlin et al., eds., *Intelligence, Genes, and Success: Scientists Respond to The Bell Curve* (New York: Springer, 1997); Ulric Neisser, ed., *The Rising Curve: Long-Term Gains in IQ and Related Measures* (Washington, D.C.: American Psychological Association, 1998); David Rowe, "A Place at the Policy Table: Behavior Genetics and Estimates of Family Environmental Effects on IQ," *Intelligence* 24 (1997): 133–159; Sternberg and Grigorenko (1997), and Christopher Jencks and Meredith Phillips, *The Black–White Test Score Gap* (Washington, D.C.: Brookings Institution Press, 1998).

31. According to this study, "Across the ordinary range of environments in modern Western societies, a sizable part of the variation in intelligence test scores is associated with genetic differences among individuals . . . If one simply combines all available correlations in a single analysis, the heritability ($h^2$) works out to about .50 . . . These overall figures are misleading, however, because most of the relevant studies have been done with children. We now know that the heritability of IQ changes with age: $h^2$ goes up and $c^2$ [the similarity in intelligence of unrelated people reared together] goes down from infancy to adulthood . . . The correlation between MZ [monozygotic] twins reared apart, which directly estimates $h^2$, ranged from .68 to .78 in five studies involving adult samples from Europe and the United States." Ulric Neisser and Gweneth Boodoo et al., "Intelligence: Knowns and Unknowns," *American Psychologist* 51 (1996): 77–101.

32. Michael Daniels, Bernie Devlin, and Kathryn Roeder, "Of Genes and IQ," in Devlin et al. (1997).

33. James Robert Flynn, "Massive IQ Gains in 14 Nations: What IQ Tests Really Measure," *Psychological Bulletin* 101 (1987): 171–191; and "The Mean IQ of Americans: Massive Gains 1932–1978," *Psychological Bulletin* 95 (1984): 29–51.

34. For an account of Lombroso's work, see James Q. Wilson and Richard J. Herrnstein, *Crime and Human Nature* (New York: Simon and Schuster, 1985), pp. 72–75.

35. Sarnoff Mednick and William Gabrielli, "Genetic Influences in Criminal Convictions: Evidence from an Adoption Cohort," *Science* 224 (1984): 891–894; and Sarnoff Mednick and Terrie E. Moffit, *The Causes of Crime: New Biological Approaches* (New York: Cambridge University Press, 1987).

36. Wilson and Herrnstein (1985), p. 94.

37. For one such critique, see Troy Duster, *Backdoor to Eugenics* (New York: Routledge, 1990), pp. 96–101.

38. Travis Hirschi and Michael Gottfredson, *A General Theory of Crime* (Stanford, Calif.: Stanford University Press, 1990).

39. H. Stattin and I. Klackenberg-Larsson, "Early Language and Intelligence Development and Their Relationship to Future Criminal Behavior," *Journal of Abnormal Psychology* 102 (1993): 369–378.

40. For a systematic account of the evidence for this, see Wilson and Herrnstein (1985), pp. 104–147.

41. Richard Wrangham and Dale Peterson, *Demonic Males: Apes and the Origins of Human Violence* (Boston: Houghton Mifflin, 1996).

42. For further examples of chimp violence, see Frans de Waal, *Chimpanzee Politics: Power and Sex among Apes* (Baltimore: Johns Hopkins University Press, 1989).

43. H. G. Brunner, "Abnormal Behavior Associated with a Point Mutation in the Structural Gene for Monoamine Oxidase A," *Science* 262 (1993): 578–580.

44. Lois Wingerson, *Unnatural Selection: The Promise and the Power of Human Gene Research* (New York: Bantam Books, 1998), pp. 291–294.

45. The theory that crime is the result of a failure to learn impulse control at a certain key developmental stage is sometimes referred to as the "life course" theory of crime; it offers an explanation as to why so large a percentage of crimes are committed by recidivists. The classic study establishing the existence of criminal "life courses" is Sheldon Glueck and Eleanor Glueck, *Delinquency and Nondelinquency in Perspective* (Cambridge, Mass.: Harvard University Press, 1968). See also the reanalysis of the Gluecks' data in Robert J. Sampson and John H. Laub, *Crime in the Making: Pathways and Turning Points Through Life* (Cambridge, Mass.: Harvard University Press, 1993).

46. For an account of the rise and fall of crime rates in the United States and other Western countries after 1965, see Francis Fukuyama, *The Great Disruption: Human Nature and the Reconstitution of Social Order* (New York: Free Press, 1999), pp. 77–87.

47. Martin Daly and Margo Wilson, *Homicide* (New York: Aldine de Gruyter, 1988).

48. For an entertaining account of this incident, see Tom Wolfe, *Hooking Up* (New York: Farrar, Straus and Giroux, 2000), pp. 92–94.

49. Wingerson (1998), pp. 294–297.

50. David Wasserman, "Science and Social Harm: Genetic Research into Crime and Violence," *Report from the Institute for Philosophy and Public Policy* 15 (1995): 14–19.

51. Wade Roush, "Conflict Marks Crime Conference; Charges of Racism and Eugenics Exploded at a Controversial Meeting," *Science* 269 (1995): 1808–1809.

52. Alice H. Eagley, "The Science and Politics of Comparing Women and Men," *American Psychologist* 50 (1995): 145–158.

53. Donald Symons, *The Evolution of Human Sexuality* (Oxford: Oxford University Press, 1979).

54. Eleanor E. Maccoby and Carol N. Jacklin, *Psychology of Sex Differences* (Stanford, Calif.: Stanford University Press, 1974).

55. Ibid., pp. 349–355.

56. Eleanor E. Maccoby, *The Two Sexes: Growing Up Apart, Coming Together* (Cambridge, Mass.: Belknap/Harvard, 1998), pp. 32–58.

57. Ibid., pp. 89–117.

58. Matt Ridley, *The Red Queen: Sex and the Evolution of Human Nature* (New York: Macmillan, 1993), pp. 279–280. Ridley cites another theory by Hurst and Haig that suggests the "gay gene" may be in the mitochondria and is similar to the "male killer" genes found in many insects.

59. Simon LeVay, "A Difference in Hypothalamic Structure Between Heterosexual and Homosexual Men," *Science* 253 (1991): 1034–1037.

60. Dean Hamer, "A Linkage Between DNA Markers on the X Chromosome and Male Sexual Orientation," *Science* 261 (1993): 321–327.

61. William Byne, "The Biological Evidence Challenged," *Scientific American* 270, no. 5 (1994): 50–55.

62. Robert Cook-Degan, *The Gene Wars: Science, Politics, and the Human Genome* (New York: W. W. Norton, 1994), p. 253.

### CHAPTER 3: NEUROPHARMACOLOGY AND THE CONTROL OF BEHAVIOR

1. Peter D. Kramer, *Listening to Prozac* (New York: Penguin Books, 1993), p. 44; see also Tom Wolfe's account in *Hooking Up* (New York: Farrar, Straus and Giroux, 2000), pp. 100–101.

2. Roger D. Masters and Michael T. McGuire, eds., *The Neurotransmitter Revolution: Serotonin, Social Behavior, and the Law* (Carbondale, Ill.: Southern Illinois University Press, 1994).

3. Ibid., p. 10.

4. Kramer (1993); and Elizabeth Wurtzel, *Prozac Nation: A Memoir* (New York: Riverhead Books, 1994).

5. Kramer (1993), pp. 1–9.

6. Joseph Glenmullen, *Prozac Backlash: Overcoming the Dangers of Prozac, Zoloft, Paxil, and Other Antidepressants with Safe, Effective Alternatives* (New York: Simon and Schuster, 2000), p. 15.

7. Irving Kirsch and Guy Sapirstein, "Listening to Prozac but Hearing Placebo: A Meta-Analysis of Antidepressant Medication," *Prevention and Treatment* 1 (1998); Larry E. Beutler, "Prozac and Placebo: There's a Pony in There Somewhere," *Prevention and Treatment* 1 (1998); and Seymour Fisher and Roger P. Greenberg, "Prescriptions for Happiness?," *Psychology Today* 28 (1995): 32–38.

8. Peter R. Breggin and Ginger Ross Breggin, *Talking Back to Prozac: What Doctors Won't Tell You About Today's Most Controversial Drug* (New York: St. Martin's Press, 1994).

9. Glenmullen (2000).

10. Robert H. Frank, *Choosing the Right Pond: Human Behavior and the Quest for Status* (Oxford: Oxford University Press, 1985).

11. For an extended discussion of the role of recognition in history, see Francis Fukuyama, *The End of History and the Last Man* (New York: Free Press, 1992), pp. 143–244.

12. Frans de Waal, *Chimpanzee Politics: Power and Sex among Apes* (Baltimore: Johns Hopkins University Press, 1989).

13. Frank (1985), pp. 21–25.

14. Related drugs include dextroamphetamine (Dexedrine), Adderall, Dextrostat, and pemoline (Cylert).

15. Dorothy Bonn, "Debate on ADHD Prevalence and Treatment Continues," *The Lancet* 354, issue 9196 (1999): 2139.

16. Edward M. Hallowell and John J. Ratey, *Driven to Distraction: Recognizing and Coping with Attention Deficit Disorder from Childhood Through Adulthood* (New York: Simon and Schuster, 1994).

17. Lawrence H. Diller, "The Run on Ritalin: Attention Deficit Disorder and Stimulant Treatment in the 1990s," *Hasting Center Report* 26 (1996): 12–18.

18. Lawrence H. Diller, *Running on Ritalin* (New York: Bantam Books, 1998), p. 63.

19. For an excellent overall treatment of the controversy over Ritalin, see Mary Eberstadt, "Why Ritalin Rules," *Policy Review*, April–May 1999, 24–44.

20. Diller (1998), p. 63.

21. Doug Hanchett, "Ritalin Speeds Way to Campuses—College Kids Using Drug to Study, Party," *Boston Herald*, May 21, 2000, p. 8.

22. Elizabeth Wurtzel, "Adventures in Ritalin," *The New York Times*, April 1, 2000, p. A15.

23. Harold S. Koplewicz, *It's Nobody's Fault: New Hope and Help for Difficult Children and Their Parents* (New York: Times Books, 1997).

24. On the politics of Ritalin, see Neil Munro, "Brain Politics," *National Journal* 33 (2001): 335–339.

25. For more on this, see the CHADD Web site, https://chadd.safeserver.com/about_chadd02.htm.

26. Eberstadt (1999).

27. Diller (1998), pp. 148–150.

28. Dyan Machan and Luisa Kroll, "An Agreeable Affliction," *Forbes*, August 12, 1996, 148.

29. Marsha Rappley, Patricia B. Mullan, et al., "Diagnosis of Attention-Deficit/Hyperactivity Disorder and Use of Psychotropic Medication in Very Young Children," *Archives of Pediatrics and Adolescent Medicine* 153 (1999): 1039–1045.

30. Julie Magno Zito, Daniel J. Safer, et al., "Trends in the Prescribing of Psychotropic Medications to Preschoolers," *Journal of the American Medical Association* 283 (2000): 1025–1060.

31. I am grateful to Michael McGuire for help with this section.

32. These statistics are taken from the National Institute on Drug Abuse's Web site, http://www.nida.nih.gov/Infofax/ecstasy.html.

33. Matthew Klam, "Experiencing Ecstasy," *The New York Times Magazine*, January 21, 2001.

### CHAPTER 4: THE PROLONGATION OF LIFE

1. See http://www.demog.berkeley.edu/~andrew/1918/figure2.html for the 1900 figures, and http://www.cia.gov/cia/publications/factbook/geos/us.html for 2000.

2. For an overview of these theories, see Michael R. Rose, *Evolutionary Biology of Aging* (New York: Oxford University Press, 1991), p. 160 ff; Caleb E. Finch and Rudolph E. Tanzi, "Genetics of Aging," *Science* 278 (1997): 407–411; S. Michal Jazwinski, "Longevity, Genes, and Aging," *Science* 273 (1996): 54–59; and David M. A. Mann, "Molecular Biology's Impact on Our Understanding of Aging," *British Medical Journal* 315 (1997): 1078–1082.

3. Michael R. Rose, "Finding the Fountain of Youth," *Technology Review* 95, no. 7 (October 1992): 64–69.

4. Nicholas Wade, "A Pill to Extend Life? Don't Dismiss the Notion Too Quickly," *The New York Times*, September 22, 2000, p. A20.

5. Tom Kirkwood, *Time of Our Lives: Why Ageing Is Neither Inevitable nor Necessary* (London: Phoenix, 1999), pp. 100–117.

6. Dwayne A. Banks and Michael Fossel, "Telomeres, Cancer, and Aging: Altering the Human Life Span," *Journal of the American Medical Association* 278 (1997): 1345–1348.

7. Nicholas Wade, "Searching for Genes to Slow the Hands of Biological Time," *The New York Times*, September 26, 2000, p. D1; Cheol-Koo Lee and Roger G. Klopp et al., "Gene Expression Profile of Aging and Its Retardation by Caloric Restriction," *Science* 285 (1999): 1390–1393.

8. Kirkwood (1999), p. 166.

9. For a sample of the discussion on stem cells, see Eric Juengst and Michael Fossel, "The Ethics of Embryonic Stem Cells—Now and Forever, Cells without End," *Journal of the American Medical Association* 284 (2000): 3180–3184; Juan de Dios Vial Correa and S. E. Mons. Elio Sgreccia, *Declaration on the Production and the Scientific and Therapeutic Use of Human Embryonic Stem Cells* (Rome: Pontifical Academy for Life, 2000); and M. J. Friedrich, "Debating Pros and Cons of Stem Cell Research," *Journal of the American Medical Association* 284, no. 6 (2000): 681–684.

10. Gabriel S. Gross, "Federally Funding Human Embryonic Stem Cell Research: An Administrative Analysis," *Wisconsin Law Review* 2000 (2000): 855–884.

11. For some research strategies into therapies for aging, see Michael R. Rose, "Aging as a Target for Genetic Engineering," in Gregory Stock and John Campbell, eds., *Engineering the Human Germline: An Exploration of the Science and Ethics of Altering the Genes We Pass to Our Children* (New York: Oxford University Press, 2000), pp. 53–56.

12. Jean Fourastié, "De la vie traditionelle à la vie tertiaire," *Population* 14 (1963): 417–432.

13. Kirkwood (1999), p. 6.

14. "Resident Population Characteristics—Percent Distribution and Median Age, 1850–1996, and Projections, 2000–2050," www.doi.gov/nrl/statAbst/Aidemo.pdt.

15. Nicholas Eberstadt, "World Population Implosion?," *Public Interest*, no. 129 (February 1997): 3–22.

16. On this issue, see Francis Fukuyama, "Women and the Evolution of World Politics," *Foreign Affairs* 77 (1998): 24–40.

17. Pamela J. Conover and Virginia Sapiro, "Gender, Feminist Consciousness, and War," *American Journal of Political Science* 37 (1993): 1079–1099.

18. Edward N. Luttwak, "Toward Post-Heroic Warfare," *Foreign Affairs* 74 (1995): 109–122.

19. For a longer discussion of this, see Francis Fukuyama, *The Great Disruption: Human Nature and the Reconstitution of Social Order* (New York: Free Press, 1999), pp. 212–230.

20. This point is made by Fred Charles Iklé, "The Deconstruction of Death," *The National Interest*, no. 62 (Winter 2000/01): 87–96.

21. Generational change is the theme, inter alia, of Arthur M. Schlesinger, Jr.'s, *Cycles of American History* (Boston: Houghton Mifflin, 1986); see also William Strauss and Neil Howe, *The Fourth Turning: An American Prophecy* (New York: Broadway Books, 1997).

22. Kirkwood (1999), pp. 131–132.

23. Michael Norman, "Living Too Long," *The New York Times Magazine*, January 14, 1996, pp. 36–38.

24. Kirkwood (1999), p. 238.

25. On the evolution of human sexuality, see Donald Symons, *The Evolution of Human Sexuality* (Oxford: Oxford University Press, 1979).

## CHAPTER 5: GENETIC ENGINEERING

1. On the history of the Human Genome Project, see Robert Cook-Degan, *The Gene Wars: Science, Politics, and the Human Genome* (New York: W. W. Norton, 1994); Kathryn Brown, "The Human Genome Business Today," *Scientific American* 283 (July 2000): 50–55; and Kevin Davies, *Cracking the Genome: Inside the Race to Unlock Human DNA* (New York: Free Press, 2001).

2. Carol Ezzell, "Beyond the Human Genome," *Scientific American* 283, no. 1 (July 2000): 64–69.

3. Ken Howard, "The Bioinformatics Gold Rush," *Scientific American* 283, no. 1 (July 2000): 58–63.

4. Interview with Stuart A. Kauffman, "Forget In Vitro—Now It's 'In Silico,' " *Scientific American* 283, no. 1 (July 2000): 62–63.

5. Gina Kolata, "Genetic Defects Detected in Embryos Just Days Old," *The New York Times*, September 24, 1992, p. A1.

6. Lee M. Silver, *Remaking Eden: Cloning and Beyond in a Brave New World* (New York: Avon, 1998), pp. 233–247.

7. Ezzell (2000).

8. For Wilmut's own account of this accomplishment, see Ian Wilmut, Keith Campbell, and Colin Tudge, *The Second Creation: Dolly and the Age of Biological Control* (New York: Farrar, Straus and Giroux, 2000).

9. National Bioethics Advisory Commission, *Cloning Human Beings* (Rockville, Md.: National Bioethics Advisory Commission, 1997).

10. Margaret Talbot, "A Desire to Duplicate," *The New York Times Magazine*, February 4, 2001, pp. 40–68; Brian Alexander, "(You)2," *Wired*, February 2001, 122–135.

11. Glenn McGee, *The Perfect Baby: A Pragmatic Approach to Genetics* (Lanham, Md.: Rowman and Littlefield, 1997).

12. For an overview of the present state of human germ-line engineering, see Gregory Stock and John Campbell, eds., *Engineering the Human Germline: An Exploration of the Science and Ethics of Altering the Genes We Pass to Our Children* (New York: Oxford University Press, 2000); Marc Lappé, "Ethical Issues in Manipulating the Human Germ Line," in Peter Singer and Helga Kuhse, eds., *Bioethics: An Anthology* (Oxford: Blackwell, 1999), p. 156; and Mark S. Frankel and Audrey R. Chapman, *Human Inheritable Genetic Modifications: Assessing Scientific, Ethical, Religious, and Policy Issues* (Washington, D.C.: American Association for the Advancement of Science, 2000).

13. On the technology of artificial chromosomes, see John Campbell and Gregory Stock, "A Vision for Practical Human Germline Engineering," in Stock and Campbell, eds. (2000), pp. 9–16.

14. Edward O. Wilson, "Reply to Fukuyama," *The National Interest*, no. 56 (Spring 1999): 35–37.

15. Gina Kolata, *Clone: The Road to Dolly and the Path Ahead* (New York: William Morrow, 1998), p. 27.

16. W. French Anderson, "A New Front in the Battle against Disease," in Stock and Campbell, eds. (2000), p. 43.

17. Fred Charles Iklé, "The Deconstruction of Death," *The National Interest*, no. 62 (Winter 2000/01): 91–92.

18. Kolata (1998), pp. 120–156.

19. Nicholas Eberstadt, "Asia Tomorrow, Gray and Male," *The National Interest* 53 (1998): 56–65, Terence H. Hull, "Recent Trends in Sex Ratios at Birth in China," *Population and Development Review* 16 (1990): 63–83; Chai Bin Park, "Preference for Sons, Family Size, and Sex Ratio: An Empirical Study in Korea," *Demography* 20 (1983): 333–352; and Barbara D. Miller, *The Endangered Sex: Neglect of Female Children in Rural Northern India* (Ithaca, N.Y., and London: Cornell University Press, 1981).

20. Elisabeth Croll, *Endangered Daughters: Discrimination and Development in Asia* (London: Routledge, 2001); and Ansley J. Coale and Judith Banister, "Five Decades of Missing Females in China," *Demography* 31 (1994): 459–479.

21. Gregory S. Kavka, "Upside Risks," in Carl F. Cranor, ed., *Are Genes Us?: Social Consequences of the New Genetics* (New Brunswick, N.J.: Rutgers University Press, 1994), p. 160.

22. This scenario has been suggested by Charles Murray. See "Deeper into the Brain," *National Review* 52 (2000): 46–49.

## CHAPTER 6: WHY WE SHOULD WORRY

1. Rifkin's extensive writings on biotechnology include *Algeny: A New Word, a New World* (New York: Viking, 1983); and, with Ted Howard, *Who Should Play God?* (New York: Dell, 1977).

2. I am grateful to Michael Lind for pointing out the roles of Haldane, Bernal, and Shaw in this regard.

3. Quoted in Diane B. Paul, *Controlling Human Heredity: 1865 to the Present* (Atlantic Highlands, N.J.: Humanities Press, 1995), p. 2. See also her article "Eugenic Anxieties, Social Realities, and Political Choices," *Social Research* 59 (1992): 663–683. See also Mark H. Haller, *Eugenics: Hereditarian Attitudes in American Thought* (New Brunswick, N.J.: Rutgers University Press, 1963).

4. See Henry P. David and Jochen Fleischhacker, "Abortion and Eugenics in Nazi Germany," *Population and Development Review* 14 (1988): 81–112.

5. The classic study of this is Robert Jay Lifton, *The Nazi Doctors: Medical Killing and the Psychology of Genocide* (New York: Basic Books, 1986).

6. Gunnar Broberg and Nils Roll-Hansen, *Eugenics and the Welfare State: Sterilization Policy in Denmark, Sweden, Norway, and Finland* (East Lansing, Mich.: Michigan State University Press, 1996). See also Mark B. Adams, *The Wellborn Science: Eugenics in Germany, France, Brazil, and Russia* (New York and Oxford: Oxford University Press, 1990).

7. For a history of eugenics in China, see Frank Dikötter, *Imperfect Conceptions: Medical Knowledge, Birth Defects and Eugenics in China* (New York: Columbia University Press, 1998). See also his article "Throw-Away Babies: The Growth of Eugenics Policies and Practices in China," *The Times Literary Supplement*, January 12, 1996, pp. 4–5; and Veronica Pearson, "Population Policy and Eugenics in China," *British Journal of Psychiatry* 167 (1995): 1–4.

8. Diane B. Paul, "Is Human Genetics Disguised Eugenics?" in David L. Hull and Michael Ruse, eds., *The Philosophy of Biology* (New York: Oxford University Press, 1998), pp. 536ff.

9. Pearson (1995), p. 2.

10. Matt Ridley, *Genome: The Autobiography of a Species in 23 Chapters* (New York: HarperCollins, 2000), pp. 297–299.

11. Robert L. Sinsheimer, "The Prospect of Designed Genetic Change," in Ruth F. Chadwick, ed., *Ethics, Reproduction, and Genetic Control*, rev. ed. (London and New York: Routledge, 1992), p. 145.

12. China's one-child population policy and the forced abortions it has entailed have been controversial among many conservative groups in the United States. See Steven Mosher, *A Mother's Ordeal: One Woman's Fight against China's One-Child Policy* (New York: Harcourt Brace Jovanovich, 1993).

13. Kate Devine, "NIH Lifts Stem Cell Funding Ban, Issues Guidelines," *The Scientist* 14, no. 18 (2000): 8.

14. Charles Krauthammer, "Why Pro-Lifers Are Missing the Point: The Debate over Fetal-Tissue Research Overlooks the Big Issue," *Time*, February 12, 2001, 60.

15. Virginia I. Postrel, *The Future and Its Enemies: The Growing Conflict over Creativity, Enterprise, and Progress* (New York: Touchstone Books, 1999), p. 168.

16. Mark K. Sears et al., "Impact of Bt Corn Pollen on Monarch Buterflies: A Risk Assessment," *Proceedings of the National Academy of Sciences* 98 (October 9, 2001): 11937–11942.

17. For an intelligent discussion of some possible negative externalities of biotechnology, see Gregory S. Kavka, "Upside Risks," in Carl F. Cranor, ed., *Are Genes Us?: Social Consequences of the New Genetics* (New Brunswick, N.J.: Rutgers University Press, 1994).

18. John Colapinto, *As Nature Made Him: The Boy Who Was Raised As a Girl* (New York: HarperCollins, 2000), p. 58.

19. Colapinto (2000), pp. 69–70.

20. Kavka, in Cranor, ed. (1994), pp. 164–165.

21. Richard D. Alexander, *How Did Humans Evolve? Reflections on the Uniquely Unique Species* (Ann Arbor, Mich.: Museum of Zoology, University of Michigan, 1990), p. 6.

22. Plato, *The Republic*, Book V, 457c–e.

23. Gary S. Becker, "Crime and Punishment: An Economic Approach," *Journal of Political Economy* 76 (1968): 169–217.

## CHAPTER 7: HUMAN RIGHTS

1. This quotation is taken from the transcript of a conference discussion reprinted in John Stock and Gregory Campbell, eds., *Engineering the Human Germline: An Exploration of the Science and Ethics of Altering the Genes We Pass to Our Children* (New York: Oxford University Press, 2000), p. 85.

2. This argument is made in Ronald M. Dworkin, *Life's Dominion: An Argument about Abortion, Euthanasia, and Individual Freedom* (New York: Vintage Books, 1994).

3. John A. Robertson, *Children of Choice: Freedom and the New Reproductive Technologies* (Princeton, N.J.: Princeton University Press, 1994), pp. 33–34.

4. Ronald M. Dworkin, *Sovereign Virtue: The Theory and Practice of Equality* (Cambridge, Mass.: Harvard University Press, 2000), p. 452. For an excellent critique, see Adam Wolfson, "Politics in a Brave New World," *Public Interest* no. 142 (Winter 2001): 31–43.

5. G. E. Moore actually coined the phrase *naturalistic fallacy*. See his *Principia Ethica* (Cambridge: Cambridge University Press, 1903), p. 10.

6. For an example of a recent statement of this, see Alexander Rosenberg, *Darwinism in Philosophy, Social Science, and Policy* (Cambridge: Cambridge University Press, 2000), p. 120.

7. Paul Ehrlich, *Human Natures: Genes, Cultures, and the Human Prospect* (Washington, D.C./Covelo, Calif.: Island Press/Shearwater Books, 2000), p. 309.

8. William F. Schultz, letter to the editor, *The National Interest*, no. 63 (Spring 2001): 124–125.

9. David Hume, *A Treatise of Human Nature*, Book III, part I, section I (London: Penguin Books, 1985), p. 521.

10. Robin Fox, "Human Nature and Human Rights," *The National Interest*, no. 62 (Winter 2000/01): 77–86.

11. Ibid., p. 78.

12. Alasdair MacIntyre, "Hume on 'Is' and 'Ought,'" *Philosophical Review* 68 (1959): 451–468.

13. This point is made in Robert J. McShea, "Human Nature Theory and Political Philosophy," *American Journal of Political Science* 22 (1978): 656–679. For a typical misunderstanding of Aristotle, see Allen Buchanan and Norman Daniels et al., *From Chance to Choice: Genetics and Justice* (New York and Cambridge: Cambridge University Press, 2000), p. 89.

14. Robert J. McShea, *Morality and Human Nature: A New Route to Ethical Theory* (Philadelphia: Temple University Press, 1990), pp. 68–69.

15. See the discussion of Bentham in Charles Taylor, *Sources of the Self: The Making of the Modern Identity* (Cambridge, Mass.: Harvard University Press, 1989), p. 332.

16. Hume has been wrongly interpreted as a kind of proto-Kantian, when in fact he falls squarely in the older tradition of deriving rights from human nature.

17. Immanuel Kant, *Foundations of the Metaphysics of Morals*, trans. Lewis White Beck (Indianapolis: Bobbs-Merrill, 1959), p. 9.

18. This includes MacIntyre (1959), pp. 467–468.

19. John Rawls, *A Theory of Justice*, rev. ed. (Cambridge, Mass.: Harvard/Belknap, 1999), p. 17.

20. Ibid., pp. 347–365.

21. William A. Galston, "Liberal Virtues," *American Political Science Review* 82, no. 4 (December 1988): 1277–1290.

22. Ackerman quoted in William A. Galston, "Defending Liberalism," *American Political Science Review* 76 (1982): 621–629.

23. See, for example, Allan Bloom, *Giants and Dwarfs: Essays 1960–1990* (New York: Simon and Schuster, 1990).

24. Rawls (1999), p. 433.

25. Dworkin (2000), p. 448.

26. Robertson (1994), p. 24.

27. *Casey v. Planned Parenthood* quoted in Hadley Arkes, "Liberalism and the Law," in Hilton Kramer and Roger Kimball, eds., *The Betrayal of Liberalism: How the Disciples of Freedom and Equality Helped Foster the Illiberal Politics of Coercion and Control* (Chicago: Ivan R. Dee, 1999), pp. 95–96. Arkes provides a good critique of this position and shows how it differs from the natural-rights perspective of the authors of the Constitution and the Bill of Rights. See also the critique of the interpretation of religious freedom as the freedom to, in effect, make up your own religion, contained in Michael J. Sandel, *Democracy's Discontent: America in Search of a Public Philosophy* (Cambridge, Mass.: Harvard University Press, 1996), pp. 55–90.

28. The degeneration of modern notions of freedom into relativism via Nietzsche and Heidegger is chronicled in Allan Bloom, *The Closing of the American Mind* (New York: Simon and Schuster, 1987).

29. For numerous examples of this, see Frans de Waal, *Chimpanzee Politics: Power and Sex among Apes* (Baltimore: Johns Hopkins University Press, 1989).

30. Francis Fukuyama, *The Great Disruption: Human Nature and the Reconstitution of Social Order* (New York: Free Press, 1999), pp. 174–175.

31. "Defensive modernization" describes a process whereby the requirements of external military competition drive internal sociopolitical organization and innovation. There are many examples of this, from reforms in post–Meiji Restoration Japan to the Internet.

32. Francis Fukuyama, "Women and the Evolution of World Politics," *Foreign Affairs* 77 (1998): 24–40.

33. Robert Wright, *Nonzero: The Logic of Human Destiny* (New York: Pantheon, 2000).

34. The intellectual landscape on the issue of group selection has recently changed somewhat with the work of biologists like David Sloan Wilson, who have made the case for multilevel (that is, both individual and group) selection. See David Sloan Wilson and Elliott Sober, *Unto Others: The Evolution and Psychology of Unselfish Behavior* (Cambridge, Mass: Harvard University Press, 1998).

35. For an overview, see Francis Fukuyama, "The Old Age of Mankind," in *The End of History and the Last Man* (New York: Free Press, 1992).

## CHAPTER 8: HUMAN NATURE

1. Paul Ehrlich, *Human Natures: Genes, Cultures, and the Human Prospect* (Washington, D.C./Covelo, Calif.: Island Press/Shearwater Books, 2000), p. 330. See Francis Fukuyama, review of Ehrlich in *Commentary*, February 2001.

2. David L. Hull, "On Human Nature," in David L. Hull and Michael Ruse, eds., *The Philosophy of Biology* (New York: Oxford University Press, 1998), p. 387.

3. Alexander Rosenberg, for example, argues that there are no "essential" characteristics of species because all species exhibit variance, and the median point of a range of variance does not constitute an essence. This is simply a semantic quibble: everyone who has written about the "nature" or "essence" of a particular species has in fact been referring to a median point of variance. Alexander Rosenberg, *Darwinism in Philosophy, Social Science, and Policy* (Cambridge: Cambridge University Press, 2000), p. 121. See also David L. Hull, "Species, Races, and Genders: Differences Are Not Deviations," in Robert F. Weir and Susan C. Lawrence, eds., *Genes, Humans, and Self-Knowledge* (Iowa City: University of Iowa Press, 1994), p. 207.

4. Michael Ruse, "Biological Species: Natural Kinds, Individuals, or What?," *British Journal for the Philosophy of Science* 38 (1987): 225–242.

5. See inter alia Richard C. Lewontin, Steven Rose, et al., *Not in Our Genes: Biology, Ideology, and Human Nature* (New York: Pantheon Books, 1984); Lewontin,

*The Doctrine of DNA: Biology as Ideology* (New York: HarperPerennial, 1992); and Lewontin, *Inside and Outside: Gene, Environment, and Organism* (Worcester, Mass.: Clark University Press, 1994).

6. Lewontin (1994), p. 25.

7. Lewontin, Rose, et al. (1984), pp. 69 ff.

8. I say "almost exclusively" because, as noted in the previous chapter, contemporary ethologists are demonstrating that certain species, like chimpanzees, are capable of passing on learning culturally and therefore exhibit a certain degree of cultural variance from one group to another.

9. See also Leon Eisenberg, "The Human Nature of Human Nature," *Science* 176 (1972): 123–128.

10. Ehrlich (2000), p. 273.

11. Aristotle, *Nicomachean Ethics* II.1, 1103a24–26.

12. Ibid., V.7, 1134b29–32.

13. See Aristotle, *Politics* I.2, 1253a29–32.

14. Roger D. Masters, "Evolutionary Biology and Political Theory," *American Political Science Review* 84 (1990): 195–210; *Beyond Relativism: Science and Human Values* (Hanover, N.H.: University Press of New England, 1993); and, with Margaret Gruter, *The Sense of Justice: Biological Foundations of Law* (Newbury Park, Calif.: Sage Publications, 1992).

15. Michael Ruse and Edward O. Wilson, "Moral Philosophy as Applied Science: A Darwinian Approach to the Foundations of Ethics," *Philosophy* 61 (1986): 173–192.

16. Larry Arnhart, *Darwinian Natural Right: The Biological Ethics of Human Nature* (Albany, N.Y.: State University of New York Press, 1998).

17. For a critique and discussion of Arnhart's views, see Richard F. Hassing, "Darwinian Natural Right?," *Interpretation* 27 (2000): 129–160; and Larry Arnhart, "Defending Darwinian Natural Right," *Interpretation* 27 (2000): 263–277.

18. Arnhart (1998), pp. 31–36.

19. Donald Brown, *Human Universals* (Philadelphia: Temple University Press, 1991), p. 77.

20. See, for example, Steven Pinker and Paul Bloom, "Natural Language and Natural Selection," *Behavioral and Brain Sciences* 13 (1990): 707–784; and Pinker, *The Language Instinct* (New York: HarperCollins, 1994).

21. For a critique, see Frans de Waal, *Chimpanzee Politics: Power and Sex among Apes* (Baltimore: Johns Hopkins University Press, 1989) pp. 57–60.

22. The argument about time was made by Benjamin Lee Whorf with regard to the Hopi, while the argument about color was a commonplace in anthropology textbooks. See Brown (1991), pp. 10–11.

23. John Locke, *An Essay Concerning Human Understanding*, Book I, chapter 3, section 7 (Amherst, N.Y.: Prometheus Books, 1995), p. 30.

24. Ibid., Book I, chapter 3, section 9, pp. 30–31.

25. Robert Trivers, "The Evolution of Reciprocal Altruism," *Quarterly Review of Biol-*

*ogy* 46 (1971): 35–56; see also Trivers, *Social Evolution* (Menlo Park, Calif.: Benjamin/Cummings, 1985).

26. Sarah B. Hrdy and Glenn Hausfater, *Infanticide: Comparative and Evolutionary Perspectives* (New York: Aldine Publishing, 1984); R. Muthulakshmi, *Female Infanticide: Its Causes and Solutions* (New Delhi: Discovery Publishing House, 1997); Lalita Panigrahi, *British Social Policy and Female Infanticide in India* (New Delhi: Munshiram Manoharlal, 1972); and Maria W. Piers, *Infanticide* (New York: W. W. Norton, 1978).

27. On this point, see Arnhart (1998), pp. 119–120.

28. If one looks at Locke's sources on infanticide, they fall into the category of the exotic travel literature that was produced in the seventeenth and eighteenth centuries to astonish Europeans with the strangeness and barbarity of foreign lands.

29. Peter Singer and Susan Reich, *Animal Liberation* (New York: New York Review Books, 1990), p. 6; and Peter Singer and Paola Cavalieri, *The Great Ape Project: Equality Beyond Humanity* (New York: St. Martin's Press, 1995).

30. This is a point originally made by Jeremy Bentham, and reiterated by Singer and Reich (1990), pp. 7–8.

31. See John Tyler Bonner, *The Evolution of Culture in Animals* (Princeton, N.J.: Princeton University Press, 1980).

32. Frans de Waal, *The Ape and the Sushi Master* (New York: Basic Books, 2001), pp. 194–202.

33. Ibid., pp. 64–65.

34. Peter Singer (in Singer and Reich, 1990) makes a bizarre argument that the case for equality is a moral idea in no way dependent on factual assertions about the actual equality of the beings involved. He argues, "There is no logically compelling reason for assuming that a factual difference in ability between two people justifies any difference in the amount of consideration we give to their needs and interests . . . " (pp. 4–5). This is plainly untrue: because children have undeveloped intellects and inadequate life experiences, we do not grant them the same freedom and regard as adults. Singer fails to address the question of where the moral idea of equality comes from, or why it should be more compelling than an alternative moral idea that seeks to rank all of natural creation hierarchically. Elsewhere he says that "the basic element—the taking into account of the interests of the being, whatever those interests may be—must, according to the principle of equality, be extended to all beings, black or white, masculine or feminine, human or nonhuman" (p. 5). Singer does not explicitly take up the question of whether we need to respect the interests of beings like flies and mosquitoes, much less viruses and bacteria. He may regard these as trivial examples, but they are not: the nature of rights depends on the nature of the species involved.

## CHAPTER 9: HUMAN DIGNITY

1. Clive Staples Lewis, *The Abolition of Man* (New York: Touchstone, 1944), p. 85.

2. Counsel of Europe, Draft Additional Protocol to the Convention on Human

Rights and Biomedicine, On the Prohibiting of Cloning Human Beings, Doc. 7884, July 16, 1997.

3. This is the theme of the second part of Francis Fukuyama, *The End of History and the Last Man* (New York: Free Press, 1992).

4. For an interpretation of this passage in Tocqueville, see Francis Fukuyama, "The March of Equality," *Journal of Democracy* 11 (2000): 11–17.

5. John Paul II, "Message to the Pontifical Academy of Sciences," October 22, 1996.

6. Daniel C. Dennett, *Darwin's Dangerous Idea: Evolution and the Meanings of Life* (New York: Simon and Schuster, 1995), pp. 35–39; see also Ernst Mayr, *One Long Argument: Charles Darwin and the Genesis of Modern Evolutionary Thought* (Cambridge, Mass.: Harvard University Press, 1991), pp. 40–42.

7. Michael Ruse and David L. Hull, *The Philosophy of Biology* (New York: Oxford University Press, 1998), p. 385.

8. Lee M. Silver, *Remaking Eden: Cloning and Beyond in a Brave New World* (New York: Avon, 1998), pp. 256–257.

9. Ruse and Hull (1998), p. 385.

10. Silver (1998), p. 277.

11. Friedrich Nietzsche, *Thus Spoke Zarathustra*, First part, section 5, from *The Portable Nietzsche*, ed. Walter Kaufmann (New York: Viking, 1968), p. 130.

12. Charles Taylor, *Sources of the Self: The Making of the Modern Identity* (Cambridge, Mass.: Harvard University Press, 1989), pp. 6–7.

13. For a fuller defense of this proposition, see Francis Fukuyama, *The Great Disruption: Human Nature and the Reconstitution of Social Order*, part II (New York: Free Press, 1999).

14. Aristotle, *Politics* I.2.13, 1254b, 16–24.

15. Ibid., I.2.18, 1255a, 22–38.

16. Ibid., I.2.19, 1255b, 3–5.

17. See, for example, Dan W. Brock, "The Human Genome Project and Human Identity," in *Genes, Humans, and Self-Knowledge*, eds. Robert F. Weir and Susan C. Lawrence et al. (Iowa City: University of Iowa Press, 1994), pp. 18–23.

18. This possibility has already been suggested by Charles Murray. See his "Deeper into the Brain," *National Review* 52 (2000): 46–49.

19. Peter Sloterdijk, "Regeln für den Menschenpark: Ein Antwortschreiben zum Brief über den Humanismus," *Die Zeit*, no. 38, September 16, 1999.

20. Jürgen Habermas, "Nicht die Natur verbietet das Klonen. Wir müssen selbst entscheiden. Eine Replik auf Dieter E. Zimmer," *Die Zeit*, no. 9, February 19, 1998.

21. For a discussion of this issue, see Allen Buchanan and Norman Daniels et al., *From Chance to Choice: Genetics and Justice* (New York and Cambridge: Cambridge University Press, 2000), pp. 17–20. See also Robert H. Blank and Masako N. Darrough, *Biological Differences and Social Equality: Implications for Social Policy* (Westport, Conn.: Greenwood Press, 1983).

22. Ronald M. Dworkin, *Sovereign Virtue: The Theory and Practice of Equality* (Cambridge, Mass.: Harvard University Press, 2000), p. 452.

23. Laurence H. Tribe, "Second Thoughts on Cloning," *The New York Times*, December 5, 1997, p. A31.

24. John Paul II (1996).

25. On the meaning of this "ontological leap," see Ernan McMullin, "Biology and the Theology of the Human," in Phillip R. Sloan, ed., *Controlling Our Desires: Historical, Philosophical, Ethical, and Theological Perspectives on the Human Genome Project* (Notre Dame, Ind.: University of Notre Dame Press, 2000), p. 367.

26. It is in fact very difficult to come up with a Darwinian explanation for the human enjoyment of music. See Steven Pinker, *How the Mind Works* (New York: W. W. Norton, 1997), pp. 528–538.

27. See, for example, Arthur Peacocke, "Relating Genetics to Theology on the Map of Scientific Knowledge," in Sloan (2000), pp. 346–350.

28. Laplace's exact words were: "We ought then to regard the present state of the universe [not just the solar system] as the effect of its anterior state and as the cause of the one which is to follow. Given an intelligence that could comprehend at one instant all the forces by which nature is animated and the respective situation of the beings who compose it—an intelligence sufficiently vast to submit these data [initial conditions] to analysis—it would embrace in the same formula the movements of the greatest bodies in the universe and those of the lightest atom; for it, nothing would be uncertain and the future, as the past, would be present to its eyes . . . The regularity which astronomy shows us in the movements of the comets doubtless exists also in all phenomena. The curve described by a simple molecule of air or vapor is regulated in a manner just as certain as the planetary orbits; the only difference between them is that which comes from our ignorance." Quoted in *Final Causality in Nature and Human Affairs*, ed. Richard F. Hassing (Washington, D.C.: Catholic University Press, 1997), p. 224.

29. Hassing, ed. (1997), pp. 224–226.

30. Peacocke, in Sloan, ed. (2000), p. 350.

31. McMullin, in Sloan, ed. (2000), p. 374.

32. On this question, see Roger D. Masters, "The Biological Nature of the State," *World Politics* 35 (1983): 161–193.

33. Andrew Goldberg and Christophe Boesch, "The Cultures of Chimpanzees," *Scientific American* 284 (2001): 60–67.

34. Larry Arnhart, *Darwinian Natural Right: The Biological Ethics of Human Nature* (Albany, N.Y.: State University of New York Press, 1998), pp. 61–62.

35. One exception to this appears to be the indigenous peoples of the American Pacific Northwest, a hunter-gatherer society that seems to have developed a state. See Robert Wright, *Nonzero: The Logic of Human Destiny* (New York: Pantheon Books, 2000), pp. 31–38.

36. Stephen Jay Gould and R. C. Lewontin, "The Spandrels of San Marco and the Panglossian Paradigm: A Critique of the Adaptionist Programme," *Proceedings of the Royal Society of London* 205 (1979): 81–98.

37. John R. Searle, *The Mystery of Consciousness* (New York: New York Review Books, 1997).

38. Daniel C. Dennett, *Consciousness Explained* (Boston: Little, Brown, 1991), p. 210.

39. John R. Searle, *The Rediscovery of the Mind* (Cambridge, Mass.: MIT Press, 1992), p. 3.

40. Hans P. Moravec, *Robot: Mere Machine to Transcendent Mind* (New York: Oxford University Press, 1999).

41. Ray Kurzweil, *The Age of Spiritual Machines: When Computers Exceed Human Intelligence* (London: Penguin Books, 2000).

42. For a critique, see Colin McGinn, "Hello HAL," *The New York Times Book Review*, January 3, 1999.

43. On this point, see Wright (2000), pp. 306–308.

44. Ibid., pp. 321–322.

45. Robert J. McShea, *Morality and Human Nature: A New Route to Ethical Theory* (Philadelphia: Temple University Press, 1990), p. 77.

46. Daniel Dennett makes the following bizarre statement in *Consciousness Explained*: "But why should it matter, you may want to ask, that a creature's desires are thwarted if they aren't conscious desires? I reply: Why would it matter more if they were conscious—especially if consciousness were a property, as some think, that forever eludes investigation? Why should a 'zombie's' crushed hopes matter less than a conscious person's crushed hopes? There is a trick with mirrors here that should be exposed and discarded. Consciousness, you say, is what matters, but then you cling to doctrines about consciousness that systematically prevent us from getting any purchase on *why* it matters" (p. 450). Dennett's question begs a more obvious one: What person in the world would care about crushing a zombie's hopes, except to the extent that the zombie was instrumentally useful to that person?

47. Jared Diamond, *The Third Chimpanzee* (New York: HarperCollins, 1992), p. 23.

48. The dualism between reason and emotion—that is, the idea that these are distinct and separable mental qualities—can be traced to Descartes (see *The Passions of the Soul*, Article 47). This dichotomy has been widely accepted since then but is misleading in many ways. The neurophysiologist Antonio Damasio points out that human reasoning invariably involves what he labels somatic markers—emotions that the mind attaches to certain ideas or options in the course of thinking through a problem—that help speed many kinds of calculations. Antonio R. Damasio, *Descartes' Error: Emotion, Reason, and the Human Brain* (New York: Putnam, 1994).

49. That is, the Kantian notion that moral choice is an act of pure reason overriding or suppressing natural emotions is not the way that human beings actually make moral choices. Human beings more typically balance one set of feelings against another and build character by strengthening the pleasurability of good moral choices through habit.

### CHAPTER 10: THE POLITICAL CONTROL OF BIOTECHNOLOGY

1. The self-interestedness of public officials is the starting premise of the Public Choice school. See James M. Buchanan and Gordon Tullock, *The Calculus of*

*Consent: Logical Foundations of Constitutional Democracy* (Ann Arbor, Mich.: University of Michigan Press, 1962); and Jack High and Clayton A. Coppin, *The Politics of Purity: Harvey Washington Wiley and the Origins of Federal Food Policy* (Ann Arbor, Mich.: University of Michigan Press, 1999).

2. Quoted in Gregory Stock and John Campbell, eds., *Engineering the Human Germline: An Exploration of the Science and Ethics of Altering the Genes We Pass to Our Children* (New York: Oxford University Press, 2000), p. 78.

3. For a general theory of where the state can legitimately intervene in family matters, see Gary S. Becker, "The Family and the State," *Journal of Law and Economics* 31 (1988): 1–18. Becker argues that the state needs to intervene only in cases where the interests of children are not adequately represented, which would seem to be the case with cloning.

4. I myself have been guilty of this kind of thinking. See Francis Fukuyama, Caroline Wagner, et al., *Information and Biological Revolutions: Global Governance Challenges—A Summary of a Study Group* (Santa Monica, Calif.: Rand MR-1139-DARPA, 1999).

5. See, for example, P.M.S. Blackett, *Fear, War, and the Bomb* (New York: McGraw-Hill, 1948).

6. Etel Solingen, "The Political Economy of Nuclear Restraint," *International Security* 19 (1994): 126–169.

7. Frans de Waal, *The Ape and the Sushi Master* (New York: Basic Books, 2001), p. 116.

8. Drugs can also be approved at the nation-state level, and across jurisdictions, under a mutual-recognition procedure.

9. Bryan L. Walser, "Shared Technical Decisionmaking and the Disaggregation of Sovereignty," *Tulane Law Review* 72 (1998): 1597–1697.

#### CHAPTER 11: HOW BIOTECHNOLOGY IS REGULATED TODAY

1. Kurt Eichenwald, "Redesigning Nature: Hard Lessons Learned; Biotechnology Food: From the Lab to a Debacle," *The New York Times*, January 25, 2001, p. A1.

2. Donald L. Uchtmann and Gerald C. Nelson, "US Regulatory Oversight of Agricultural and Food-Related Biotechnology," *American Behavioral Scientist* 44 (2000): 350–377.

3. Uchtmann and Nelson (2000), and Sarah E. Taylor, "FDA Approval Process Ensures Biotech Safety," *Journal of the American Dietetic Association* 100, no. 10 (2000): 3.

4. There are, nonetheless, criticisms of excessive biotech regulation, particularly on the part of the Environmental Protection Agency. See Henry I. Miller, "A Need to Reinvent Biotechnology Regulation at the EPA," *Science* 266 (1994): 1815–1819.

5. Alan McHughen, *Pandora's Picnic Basket: The Potential and Hazards of Genetically Modified Foods* (Oxford: Oxford University Press, 2000), pp. 149–152.

6. Lee Ann Patterson, "Biotechnology Policy: Regulating Risks and Risking Regula-

tion," in Helen Wallace and William Wallace, eds., *Policy-Making in the European Union* (Oxford and New York: Oxford University Press, 2000), pp. 321–323.

7. Technically, an importer wanting to market a GMO in Europe must first apply to the competent authority in the member state into which the product is first to be marketed. If the member state gives its approval, the dossier of information is then forwarded to the commission in Brussels, which circulates it to all the other member states for comment. If none of the other member states objects, then the product can be marketed throughout the EU. In 1997, Austria and Luxemburg initiated bans on the import and cultivation of insect-resistant corn, which the commission required they rescind.

   See Ruth MacKenzie and Silvia Francescon, "The Regulation of Genetically Modified Foods in the European Union: An Overview," *N.Y.U. Environmental Law Journal* 8 (2000): 530–554.

8. Margaret R. Grossman and A. Bryan, "Regulation of Genetically Modified Organisms in the European Union," *American Behavioral Scientist* 44 (2000): 378–434; and Marsha Echols, "Food Safety Regulation in the EU and the US: Different Cultures, Different Laws," *Columbia Journal of European Law* 23 (1998): 525–543.

9. The 1990 directives do not mention the precautionary principle, but their language is not inconsistent with it. The first explicit mention of the precautionary principle is made in the Maastricht Treaty of 1992. See MacKenzie and Francescon (2000). See also Jonathan H. Adler, "More Sorry Than Safe: Assessing the Precautionary Principle and the Proposed International Biosafety Protocol," *Texas International Law Journal* 35, no. 2 (2000): 173–206.

10. Patterson, in Wallace and Wallace (2000), pp. 324–328.

11. World Trade Organization, *Trading into the Future*, 2d ed., rev. (Lausanne: World Trade Organization, 1999), p. 19.

12. Lewis Rosman, "Public Participation in International Pesticide Regulation: When the Codex Commission Decides," *Virginia Environmental Law Journal* 12 (1993): 329.

13. Aarti Gupta, "Governing Trade in Genetically Modified Organisms: The Cartagena Protocol on Biosafety," *Environment* 42 (2000): 22–27.

14. Kal Raustiala and David Victor, "Biodiversity since Rio: The Future of the Convention on Biological Diversity," *Environment* 38 (1996): 16–30.

15. Robert Paarlberg, "The Global Food Fight," *Foreign Affairs* 79 (2000): 24–38; and Nuffield Council on Bioethics, *Genetically Modified Crops: The Ethical and Social Issues* (London: Nuffield Council on Bioethics, 1999).

16. Henry I. Miller and Gregory Conko, "The Science of Biotechnology Meets the Politics of Global Regulation," *Issues in Science and Technology* 17 (2000): 47–54.

17. Henry I. Miller, "A Rational Approach to Labeling Biotech-Derived Foods," *Science* 284 (1999): 1471–1472; and Alexander G. Haslberger, "Monitoring and Labeling for Genetically Modified Products," *Science* 287 (2000): 431–432.

18. Michelle D. Miller, "The Informed-Consent Policy of the International Conference on Harmonization of Technical Requirements for Registration of

Pharmaceuticals for Human Use: Knowledge Is the Best Medicine," *Cornell International Law Journal* 30 (1997): 203–244.

19. Paul M. McNeill, *The Ethics and Politics of Human Experimentation* (Cambridge: Cambridge University Press, 1993), pp. 54–55.

20. Ibid., pp. 57, 61.

21. Ibid., pp. 62–63.

22. National Bioethics Advisory Commission, *Ethical and Policy Issues in Research Involving Human Participants, Final Recommendations* (Rockville, Md.: 2001). See http://bioethics.gov/press/finalrecomm5-18.html.

23. Michele D. Miller (1997); McNeill (1993), pp. 42–43.

24. The standard work on this subject is Robert Jay Lifton, *The Nazi Doctors: Medical Killing and the Psychology of Genocide* (New York: Basic Books, 1986).

25. The Nuremberg Code was a case in which international law drove national practice, rather than the other way around, as is more usual. The American Medical Association, for example, did not formulate its own rules for human medical experimentation until after the Nuremberg Code was adopted. See Michele D. Miller (1997), p. 211.

26. McNeill (1993), pp. 44–46.

## CHAPTER 12: POLICIES FOR THE FUTURE

1. David Firn, "Biotech Industry Plays Down UK Cloning Ruling," *Financial Times*, November 15, 2001.

2. Noelle Lenoir, "Europe Confronts the Embryonic Stem Cell Research Challenge," *Science* 287 (2000): 1425–1426; and Rory Watson, "EU Institutions Divided on Therapeutic Cloning," *British Medical Journal* 321 (2000): 658.

3. Sherylynn Fiandaca, "In Vitro Fertilization and Embryos: The Need for International Guidelines," *Albany Law Journal of Science and Technology* 8 (1998): 337–404.

4. Dorothy Nelkin and Emily Marden, "Cloning: A Business without Regulation," *Hofstra Law Review* 27 (1999): 569–578.

5. For the fullest explication of this case, see Leon Kass, "Preventing a Brave New World: Why We Should Ban Cloning Now," *The New Republic*, May 21, 2001, pp. 30–39; see also Sophia Kolehmainen, "Human Cloning: Brave New Mistake," *Hofstra Law Review* 27 (1999): 557–568; and Vernon J. Ehlers, "The Case Against Human Cloning," *Hofstra Law Review* 27 (1999): 523–532; Dena S. Davis, "Religious Attitudes towards Cloning: A Tale of Two Creatures," *Hofstra Law Review* 27 (1999): 569–578; Leon Eisenberg, "Would Cloned Human Beings Really Be Like Sheep?," *New England Journal of Medicine* 340 (1999): 471–475; Eric A. Posner and Richard A. Posner, "The Demand for Human Cloning," *Hofstra Law Review* 27 (1999): 579–608; and Harold T. Shapiro, "Ethical and Policy Issues of Human Cloning," *Science* 277 (1997): 195–197. See also the different perspectives in Glenn McGee, *The Human Cloning Debate* (Berkeley, Calif.: Berkeley Hills Books, 1998).

6. See also Francis Fukuyama, "Testimony Before the Subcommittee on Health, Committee on Energy and Commerce, Regarding H.R. 1644, 'The Human Cloning Prohibition Act of 2001,' and H.R. 2172, 'The Cloning Prohibition Act of 2001,'" June 20, 2001.

7. Michel Foucault, *Madness and Civilization: A History of Insanity in the Age of Reason* (New York: Pantheon Books, 1965).

8. The biotech firm Genentech has in fact been accused of trying to push the envelope for use of its growth hormone on children who are short but not hormonally deficient. See Tom Wilke, *Perilous Knowledge: The Human Genome Project and Its Implications* (Berkeley and Los Angeles: University of California Press, 1993), pp. 136–139.

9. Lee M. Silver, *Remaking Eden: Cloning and Beyond in a Brave New World* (New York: Avon, 1998), p. 268.

10. Leon Kass, *Toward a More Natural Science: Biology and Human Affairs* (New York: Free Press, 1985), p. 173.

11. On this general topic, see James Q. Wilson, *Bureaucracy: What Government Agencies Do and Why They Do It* (New York: Basic Books, 1989).

12. Eugene Russo, "Reconsidering Asilomar," *The Scientist* 14 (April 3, 2000): 15–21; and Marcia Barinaga, "Asilomar Revisited: Lessons for Today?," *Science* 287 (March 3, 2000): 1584–1585.

13. Stuart Auchincloss, "Does Genetic Engineering Need Genetic Engineers?," *Boston College Environmental Affairs Law Review* 20 (1993): 37–64.

14. Kurt Eichenwald, "Redesigning Nature: Hard Lessons Learned; Biotechnology Food: From the Lab to a Debacle," *The New York Times*, January 25, 2001, p. A1.

# BIBLIOGRAPHY

Ackerman, Bruce. *Social Justice in the Liberal State*. New Haven, Conn.: Yale University Press, 1980.

Adams, Mark B. *The Wellborn Science: Eugenics in Germany, France, Brazil, and Russia*. New York and Oxford: Oxford University Press, 1990.

Adler, Jonathan H. "More Sorry Than Safe: Assessing the Precautionary Principle and the Proposed International Biosafety Protocol." *Texas International Law Journal* 35, no. 2 (2000): 173–206.

Alexander, Brian. "(You)2." *Wired*, February 2001: 122–135.

Alexander, Richard D. *How Did Humans Evolve? Reflections on the Uniquely Unique Species*. Ann Arbor, Mich.: Museum of Zoology, University of Michigan, 1990.

Aristotle. *Nicomachean Ethics*.

——. *Politics*.

Arnhart, Larry. *Darwinian Natural Right: The Biological Ethics of Human Nature*. Albany, N.Y.: State University of New York Press, 1998.

———. "Defending Darwinian Natural Right." *Interpretation* 27 (2000): 263–277.

Auchincloss, Stuart. "Does Genetic Engineering Need Genetic Engineers?" *Boston College Environmental Affairs Law Review* 20 (1993): 37–64.

Bacon, Sir Francis. *The Great Instauration and the Novum Organum.* Kila, Mont.: Kessinger Publishing LLC, 1997.

Banks, Dwayne A., and Michael Fossel. "Telomeres, Cancer, and Aging: Altering the Human Life Span." *Journal of the American Medical Association* 278 (1997): 1345–1348.

Barinaga, Marcia. "Asilomar Revisited: Lessons for Today?" *Science* 287 (March 3, 2000): 1584–1585.

Becker, Gary S. "Crime and Punishment: An Economic Approach." *Journal of Political Economy* 76 (1968): 169–217.

Beutler, Larry E. "Prozac and Placebo: There's a Pony in There Somewhere." *Prevention and Treatment* 1 (1998).

Blackett, P.M.S. *Fear, War, and the Bomb.* New York: McGraw-Hill, 1948.

Blank, Robert H., and Masako N. Darrough. *Biological Differences and Social Equality: Implications for Social Policy.* Westport, Conn.: Greenwood Press, 1983.

Bloom, Allan. *The Closing of the American Mind.* New York: Simon and Schuster, 1990.

———. *Giants and Dwarfs: Essays 1960–1990.* New York: Simon and Schuster, 1987.

Bonn, Dorothy. "Debate on ADHD Prevalence and Treatment Continues." *The Lancet* 354, issue 9196 (1999): 2139.

Bonner, John Tyler. *The Evolution of Culture in Animals.* Princeton, N.J.: Princeton University Press, 1980.

Bouchard, Thomas J., Jr., David T. Kykken, et al. "Sources of Human Psychological Differences: The Minnesota Study of Twins Reared Apart." *Science* 226 (1990): 223–250.

Breggin, Peter R., and Ginger Ross Breggin. *Talking Back to Prozac: What Doctors Won't Tell You About Today's Most Controversial Drug.* New York: St. Martin's Press, 1994.

Brigham, Carl C. *A Study of American Intelligence.* Princeton, N.J.: Princeton University Press, 1923.

Broberg, Gunnar, and Nils Roll-Hansen. *Eugenics and the Welfare State: Sterilization Policy in Denmark, Sweden, Norway, and Finland.* East Lansing, Mich.: Michigan State University Press, 1996.

Brown, Donald. *Human Universals.* Philadelphia: Temple University Press, 1991.

Brown, Kathryn. "The Human Genome Business Today." *Scientific American* 283, no. 1 (July 2000): 50–55.

Brunner, H. G. "Abnormal Behavior Associated with a Point Mutation in the Structural Gene for Monoamine Oxidase A." *Science* 262 (1993): 578–580.

Buchanan, Allen, Norman Daniels, et al. *From Chance to Choice: Genetics and Justice.* New York and Cambridge: Cambridge University Press, 2000.

Buchanan, James M., and Gordon Tullock. *The Calculus of Consent: Logical Founda-*

*tions of Constitutional Democracy.* Ann Arbor, Mich.: University of Michigan Press, 1962.

Byne, William. "The Biological Evidence Challenged." *Scientific American* 270, no. 5 (1994): 50–55.

Cavalli-Sforza, Luigi Luca. *Genes, Peoples, and Languages.* New York: North Point Press, 2000.

Cavalli-Sforza, Luigi Luca, and Francesco Cavalli-Sforza. *The Great Human Diasporas: The History of Diversity and Evolution.* Reading, Mass.: Addison-Wesley, 1995.

Chadwick, Ruth F., ed. *Ethics, Reproduction, and Genetic Control.* Rev. ed. London and New York: Routledge, 1992.

Cloninger, C., and M. Bohman, et al. "Inheritance of Alcohol Abuse: Crossfostering Analysis of Alcoholic Men." *Archives of General Psychiatry* 38 (1981): 861–868.

Coale, Ansley J., and Judith Banister. "Five Decades of Missing Females in China." *Demography* 31 (1994): 459–479.

Colapinto, John. *As Nature Made Him: The Boy Who Was Raised As a Girl.* New York: HarperCollins, 2000.

Conover, Pamela J., and Virginia Sapiro. "Gender, Feminist Consciousness, and War." *American Journal of Political Science* 37 (1993): 1079–1099.

Cook-Degan, Robert. *The Gene Wars: Science, Politics, and the Human Genome.* New York: W. W. Norton, 1994.

Correa, Juan de Dios Vial, and S. E. Mons. Elio Sgreccia. *Declaration on the Production and the Scientific and Therapeutic Use of Human Embryonic Stem Cells.* Rome: Pontifical Academy for Life, 2000.

Council of Europe. *Medically Assisted Procreation and the Protection of the Human Embryo: Comparative Study of 39 States.* Strasbourg: Council of Europe, 1997.

———. "On the Prohibiting of Cloning Human Beings." Draft Additional Protocol to the Convention on Human Rights and Biomedicine, Doc. 7884 (July 16, 1997).

Cranor, Carl F., ed. *Are Genes Us?: Social Consequences of the New Genetics.* New Brunswick, N.J.: Rutgers University Press, 1994.

Croll, Elisabeth. *Endangered Daughters: Discrimination and Development in Asia.* London: Routledge, 2001.

Daly, Martin, and Margo Wilson. *Homicide.* New York: Aldine de Gruyter, 1988.

Damasio, Antonio R. *Descartes' Error: Emotion, Reason, and the Human Brain.* New York: Putnam, 1994.

David, Henry P., Jochen Fleischhacker, et al. "Abortion and Eugenics in Nazi Germany." *Population and Development Review* 14 (1988): 81–112.

Davies, Kevin. *Cracking the Genome: Inside the Race to Unlock Human DNA.* New York: Free Press, 2001.

Davis, Dena S. "Religious Attitudes towards Cloning: A Tale of Two Creatures." *Hofstra Law Review* 27 (1999): 569–578.

Dennett, Daniel C. *Consciousness Explained.* Boston: Little, Brown, 1991.

———. *Darwin's Dangerous Idea: Evolution and the Meanings of Life.* New York: Simon and Schuster, 1995.

Devine, Kate. "NIH Lifts Stem Cell Funding Ban, Issues Guidelines." *The Scientist* 14, no. 18 (2000): 8.

Devlin, Bernie, et al., eds. *Intelligence, Genes, and Success: Scientists Respond to the Bell Curve*. New York: Springer, 1997.

de Waal, Frans. *The Ape and the Sushi Master*. New York: Basic Books, 2001.

———. *Chimpanzee Politics: Power and Sex among Apes*. Baltimore: Johns Hopkins University Press, 1989.

———. "The End of Nature versus Nurture." *Scientific American* 281 (1999): 56–61.

Diamond, Jared. *The Third Chimpanzee*. New York: HarperCollins, 1992.

Dikötter, Frank. *Imperfect Conceptions: Medical Knowledge, Birth Defects and Eugenics in China*. New York: Columbia University Press, 1998.

———. "Throw-Away Babies: The Growth of Eugenics Policies and Practices in China." *The Times Literary Supplement*, January 12, 1996, pp. 4–5.

Diller, Lawrence H. *Running on Ritalin*. New York: Bantam Books, 1998.

———. "The Run on Ritalin: Attention Deficit Disorder and Stimulant Treatment in the 1990s." *Hasting Center Report* 26 (1996): 12–18.

Duster, Troy. *Backdoor to Eugenics*. New York: Routledge, 1990.

Dworkin, Ronald M. *Life's Dominion: An Argument about Abortion, Euthanasia, and Individual Freedom*. New York: Vintage Books, 1994.

———. *Sovereign Virtue: The Theory and Practice of Equality*. Cambridge, Mass.: Harvard University Press, 2000.

Eagley, Alice H. "The Science and Politics of Comparing Women and Men." *American Psychologist* 50 (1995): 145–158.

Eberstadt, Mary. "Why Ritalin Rules." *Policy Review*, April–May 1999, 24–44.

Eberstadt, Nicholas. "Asia Tomorrow, Gray and Male." *The National Interest* 53 (1998): 56–65.

———. "World Population Implosion?" *Public Interest*, no. 126 (February 1997): 3–22.

Echols, Marsha. "Food Safety Regulation in the EU and the US: Different Cultures, Different Laws." *Columbia Journal of European Law* 23 (1998): 525–543.

Ehlers, Vernon J. "The Case Against Human Cloning." *Hofstra Law Review* 27 (1999): 523–532.

Ehrlich, Paul. *Human Natures: Genes, Cultures, and the Human Prospect*. Washington, D.C./Covelo, Calif.: Island Press/Shearwater Books, 2000.

Eichenwald, Kurt. "Redesigning Nature: Hard Lessons Learned; Biotechnology Food: From the Lab to a Debacle." *The New York Times*, January 25, 2001, p. A1.

Eisenberg, Leon. "The Human Nature of Human Nature." *Science* 176 (1972): 123–128.

———. "Would Cloned Human Beings Really Be Like Sheep?" *New England Journal of Medicine* 340 (1999): 471–475.

Ezzell, Carol. "Beyond the Human Genome." *Scientific American* 283, no. 1 (July 2000): 64–69.

Farmer, Anne, and Michael J. Owen. "Genomics: The Next Psychiatric Revolution?" *British Journal of Psychiatry* 169 (1996): 135–138.

Fears, Robin, Derek Roberts, et al. "Rational or Rationed Medicine? The Promise of Genetics for Improved Clinical Practice." *British Medical Journal* 320 (2000): 933–935.

Fiandaca, Sherylynn. "In Vitro Fertilization and Embryos: The Need for International Guidelines." *Albany Law Journal of Science and Technology* 8 (1998): 337–404.

Finch, Caleb E., and Rudolph E. Tanzi. "Genetics of Aging." *Science* 278 (1997): 407–411.

Fischer, Claude S., et al. *Inequality by Design: Cracking the Bell Curve Myth*. Princeton, N.J.: Princeton University Press, 1996.

Fisher, Seymour, and Roger P. Greenberg. "Prescriptions for Happiness?" *Psychology Today* 28 (1995): 32–38.

Fletcher, R. "Intelligence, Equality, Character, and Education." *Intelligence* 15 (1991): 139–149.

Flynn, James Robert. "Massive IQ Gains in 14 Nations: What IQ Tests Really Measure." *Psychological Bulletin* 101 (1987): 171–191.

———. "The Mean IQ of Americans: Massive Gains 1932–1978." *Psychological Bulletin* 95 (1984): 29–51.

Foucault, Michel. *Madness and Civilization: A History of Insanity in the Age of Reason*. New York: Pantheon Books, 1965.

Fourastié, Jean. "De la vie traditionelle à la vie tertiaire." *Population* 14 (1963): 417–432.

Fox, Robin. "Human Nature and Human Rights." *The National Interest*, no. 62 (Winter 2000/01): 77–86.

Frank, Robert H. *Choosing the Right Pond: Human Behavior and the Quest for Status*. Oxford: Oxford University Press, 1985.

Frankel, Mark S., and Audrey R. Chapman. *Human Inheritable Genetic Modifications: Assessing Scientific, Ethical, Religious, and Policy Issues*. Washington, D.C.: American Association for the Advancement of Science, 2000.

Friedrich, M. J. "Debating Pros and Cons of Stem Cell Research." *Journal of the American Medical Association* 284, no. 6 (2000): 681–684.

Fukuyama, Francis. *The End of History and the Last Man*. New York: Free Press, 1992.

———. *The Great Disruption: Human Nature and the Reconstitution of Social Order*. New York: Free Press, 1999.

———. "Is It All in the Genes?" *Commentary* 104 (September 1997): 30–35.

———. "The March of Equality." *Journal of Democracy* 11 (2000): 11–17.

———. "Second Thoughts: The Last Man in a Bottle." *The National Interest*, no. 56 (Summer 1999): 16–33.

———. "Testimony Before the Subcommittee on Health, Committee on Energy and Commerce, Regarding H.R. 1644, 'The Human Cloning Prohibition Act of 2001,' and H.R. 2172, 'The Cloning Prohibition Act of 2001.'" June 20, 2001.

———. "Women and the Evolution of World Politics." *Foreign Affairs* 77 (1998): 24–40.

Fukuyama, Francis, Caroline Wagner, et al. *Information and Biological Revolutions: Global Governance Challenges—A Summary of a Study Group*. Santa Monica, Calif.: Rand MR-1139-DARPA, 1999.

Galston, William A. "Defending Liberalism." *American Political Science Review* 76 (1982): 621–629.

———. "Liberal Virtues." *American Political Science Review* 82, no. 4 (December 1988): 1277–1290.

Galton, Francis. *Hereditary Genius: An Inquiry into Its Laws and Consequences.* New York: Appleton, 1869.

Gardner, Howard. *Frames of Mind: The Theory of Multiple Intelligences.* New York: Basic Books, 1983.

———. *Multiple Intelligences: The Theory in Practice.* New York: Basic Books, 1993.

Glenmullen, Joseph. *Prozac Backlash: Overcoming the Dangers of Prozac, Zoloft, Paxil, and Other Antidepressants with Safe, Effective Alternatives.* New York: Simon and Schuster, 2000.

Glueck, Sheldon, and Eleanor Glueck. *Delinquency and Nondelinquency in Perspective.* Cambridge, Mass.: Harvard University Press, 1968.

Goldberg, Andrew, and Christophe Boesch. "The Cultures of Chimpanzees." *Scientific American* 284 (2001): 60–67.

Gould, Stephen Jay. *The Mismeasure of Man.* New York: W. W. Norton, 1981.

Gould, Stephen Jay, and R. C. Lewontin. "The Spandrels of San Marco and the Panglossian Paradigm: A Critique of the Adaptionist Programme." *Proceedings of the Royal Society of London* 205 (1979): 81–98.

Grant, Madison. *The Passing of the Great Race; or, the Racial Basis of European History.* 4th ed., rev. New York: Charles Scribner's Sons, 1921.

Gross, Gabriel S. "Federally Funding Human Embryonic Stem Cell Research: An Administrative Analysis." *Wisconsin Law Review* 2000, no. 4 (2000): 855–884.

Grossman, Margaret R., and A. Bryan. "Regulation of Genetically Modified Organisms in the European Union." *American Behavioral Scientist* 44 (2000): 378–434.

Gupta, Aarti. "Governing Trade in Genetically Modified Organisms: The Cartagena Protocol on Biosafety." *Environment* 42 (2000): 22–27.

Guttentag, Marcia, and Paul F. Secord. *Too Many Women? The Sex Ratio Question.* Newbury Park, Calif.: Sage Publications, 1983.

Habermas, Jürgen. "Nicht die Natur verbietet das Klonen. Wir müssen selbst entscheiden. Eine Replik auf Dieter E. Zimmer." *Die Zeit,* no. 9, February 19, 1998.

Haller, Mark H. *Eugenics: Hereditarian Attitudes in American Thought.* New Brunswick, N.J.: Rutgers University Press, 1963.

Hallowell, Edward M., and John J. Ratey. *Driven to Distraction: Recognizing and Coping with Attention Deficit Disorder from Childhood Through Adulthood.* New York: Simon and Schuster, 1994.

Hamer, Dean. "A Linkage Between DNA Markers on the X Chromosome and Male Sexual Orientation." *Science* 261 (1993): 321–327.

Hanchett, Doug. "Ritalin Speeds Way to Campuses—College Kids Using Drug to Study, Party." *Boston Herald,* May 21, 2000, p. 8.

Haslberger, Alexander G. "Monitoring and Labeling for Genetically Modified Products." *Science* 287 (2000): 431–432.

Hassing, Richard F. "Darwinian Natural Right?" *Interpretation* 27 (2000): 129–160.

———, ed. *Final Causality in Nature and Human Affairs.* Washington, D.C.: Catholic University Press, 1997.

Heidegger, Martin. *Basic Writings.* New York: Harper and Row, 1957.

High, Jack, and Clayton A. Coppin. *The Politics of Purity: Harvey Washington Wiley and the Origins of Federal Food Policy.* Ann Arbor, Mich.: University of Michigan Press, 1999.

Hirschi, Travis, and Michael Gottfredson. *A General Theory of Crime.* Stanford, Calif.: Stanford University Press, 1990.

Howard, Ken. "The Bioinformatics Gold Rush." *Scientific American* 283, no. 1 (July 2000): 58–63.

Hrdy, Sarah B., and Glenn Hausfater. *Infanticide: Comparative and Evolutionary Perspectives.* New York: Aldine Publishing, 1984.

Hubbard, Ruth. *The Politics of Women's Biology.* New Brunswick, N.J.: Rutgers University Press, 1990.

Huber, Peter. *Orwell's Revenge: The 1984 Palimpsest.* New York: Free Press, 1994.

Hull, Terence H. "Recent Trends in Sex Ratios at Birth in China." *Population and Development Review* 16 (1990): 63–83.

Hume, David. *A Treatise of Human Nature.* London: Penguin Books, 1985.

Huxley, Aldous. *Brave New World.* New York: Perennial Classics, 1998.

Iklé, Fred Charles. "The Deconstruction of Death." *The National Interest,* no. 62 (Winter 2000/01): 87–96.

Jazwinski, S. Michal. "Longevity, Genes, and Aging." *Science* 273 (1996): 54–59.

Jefferson, Thomas. *The Life and Selected Writings of Thomas Jefferson.* New York: Modern Library, 1944.

Jencks, Christopher, and Meredith Phillips. *The Black–White Test Score Gap.* Washington, D.C.: Brookings Institution Press, 1998.

Jensen, Arthur R. "How Much Can We Boost IQ and Scholastic Achievement?" *Harvard Educational Review* 39 (1969): 1–123.

John Paul II. "Message to the Pontifical Academy of Sciences." October 22, 1996.

Joy, Bill. "Why the Future Doesn't Need Us." *Wired* 8 (2000): 238–246.

Joynson, Robert B. *The Burt Affair.* London: Routledge, 1989.

Juengst, Eric, and Michael Fossel. "The Ethics of Embryonic Stem Cells—Now and Forever, Cells Without End." *Journal of the American Medical Association* 284 (2000): 3180–3184.

Kamin, Leon. *The Science and Politics of IQ.* Potomac, Md.: L. Erlbaum Associates, 1974.

Kant, Immanuel. *Foundations of the Metaphysics of Morals.* Trans. Lewis White Beck. Indianapolis: Bobbs-Merrill, 1959.

Kass, Leon. "The Moral Meaning of Genetic Technology," *Commentary* 108 (1999): 32–38.

———. "Preventing a Brave New World: Why We Should Ban Cloning Now." *The New Republic,* May 21, 2001, pp. 30–39.

————. *Toward a More Natural Science: Biology and Human Affairs*. New York: Free Press, 1985.

Kevles, Daniel T., and Leroy Hood, eds. *The Code of Codes: Scientific and Social Issues in the Human Genome Project*. Cambridge, Mass.: Harvard University Press, 1992.

Kirkwood, Tom. *Time of Our Lives: Why Ageing Is Neither Inevitable nor Necessary*. London: Phoenix, 1999.

Kirsch, Irving, and Guy Sapirstein. "Listening to Prozac but Hearing Placebo: A Meta-Analysis of Antidepressant Medication." *Prevention and Treatment* 1 (1998).

Klam, Matthew. "Experiencing Ecstasy." *The New York Times Magazine*, January 21, 2001.

Kolata, Gina. *Clone: The Road to Dolly and the Path Ahead*. New York: William Morrow, 1998.

————. "Genetic Defects Detected In Embryos Just Days Old." *The New York Times*, September 24, 1992, p. A1.

Kolehmainen, Sophia. "Human Cloning: Brave New Mistake." *Hofstra Law Review* 27 (1999): 557–568.

Koplewicz, Harold S. *It's Nobody's Fault: New Hope and Help for Difficult Children and Their Parents*. New York: Times Books, 1997.

Kramer, Hilton, and Roger Kimball, eds. *The Betrayal of Liberalism: How the Disciples of Freedom and Equality Helped Foster the Illiberal Politics of Coercion and Control*. Chicago: Ivan R. Dee, 1999.

Kramer, Peter D. *Listening to Prozac*. New York: Penguin Books, 1993.

Krauthammer, Charles. "Why Pro-Lifers Are Missing the Point: The Debate over Fetal-Tissue Research Overlooks the Big Issue." *Time*, February 12, 2001, p. 60.

Kurzweil, Ray. *The Age of Spiritual Machines: When Computers Exceed Human Intelligence*. London: Penguin Books, 2000.

Lee, Cheol-Koo, Roger G. Klopp, et al. "Gene Expression Profile of Aging and Its Retardation by Caloric Restriction." *Science* 285 (1999): 1390–1393.

Lemann, Nicholas. *The Big Test: The Secret History of the American Meritocracy*. New York: Farrar, Straus and Giroux, 1999.

Lenoir, Noelle. "Europe Confronts the Embryonic Stem Cell Research Challenge." *Science* 287 (2000): 1425–1426.

LeVay, Simon. "A Difference in Hypothalamic Structure Between Heterosexual and Homosexual Men." *Science* 253 (1991): 1034–1037.

Lewis, Clive Staples. *The Abolition of Man*. New York: Touchstone, 1944.

Lewontin, Richard C. *The Doctrine of DNA: Biology as Ideology*. New York: Harper-Perennial, 1992.

————. *Inside and Outside: Gene, Environment, and Organism*. Worcester, Mass.: Clark University Press, 1994.

Lewontin, Richard C., Steven Rose, et al. *Not in Our Genes: Biology, Ideology, and Human Nature*. New York: Pantheon Books, 1984.

Lifton, Robert Jay. *The Nazi Doctors: Medical Killing and the Psychology of Genocide*. New York: Basic Books, 1986.

Locke, John. *An Essay Concerning Human Understanding*. Amherst, N.Y.: Prometheus Books, 1995.

Luttwak, Edward N. "Toward Post-Heroic Warfare." *Foreign Affairs* 74 (1995): 109–122.

Maccoby, Eleanor E. *The Two Sexes: Growing Up Apart, Coming Together*. Cambridge, Mass.: Belknap/Harvard, 1998.

Maccoby, Eleanor E., and Carol N. Jacklin. *Psychology of Sex Differences*. Stanford, Calif.: Stanford University Press, 1974.

Machan, Dyan, and Luisa Kroll. "An Agreeable Affliction." *Forbes*, August 12, 1996, 148.

MacIntyre, Alasdair. "Hume on 'Is' and 'Ought.'" *Philosophical Review* 68 (1959): 451–468.

MacKenzie, Ruth, and Silvia Francescon. "The Regulation of Genetically Modified Foods in the European Union: An Overview." *N.Y.U. Environmental Law Journal* 8 (2000): 530–554.

Mann, David M.A. "Molecular Biology's Impact on Our Understanding of Aging." *British Medical Journal* 315 (1997): 1078–1082.

Masters, Roger D. *Beyond Relativism: Science and Human Values*. Hanover, N.H.: University Press of New England, 1993.

———. "The Biological Nature of the State." *World Politics* 35 (1983): 161–193.

———. "Evolutionary Biology and Political Theory." *American Political Science Review* 84 (1990): 195–210.

Masters, Roger D., and Margaret Gruter, eds. *The Sense of Justice: Biological Foundations of Law*. Newbury Park, Calif.: Sage Publications, 1992.

Masters, Roger D., and Michael T. McGuire, eds. *The Neurotransmitter Revolution: Serotonin, Social Behavior, and the Law*. Carbondale, Ill.: Southern Illinois University Press, 1994.

Mayr, Ernst. *One Long Argument: Charles Darwin and the Genesis of Modern Evolutionary Thought*. Cambridge, Mass.: Harvard University Press, 1991.

McGee, Glenn. *The Human Cloning Debate*. Berkeley, Calif.: Berkeley Hills Books, 1998.

———. *The Perfect Baby: A Pragmatic Approach to Genetics*. Lanham, Md.: Rowman and Littlefield, 1997.

McGinn, Colin. "Hello HAL." *The New York Times Book Review*, January 3, 1999.

McHughen, Alan. *Pandora's Picnic Basket: The Potential and Hazards of Genetically Modified Foods*. Oxford: Oxford University Press, 2000.

McNeill, Paul M. *The Ethics and Politics of Human Experimentation*. Cambridge: Cambridge University Press, 1993.

McShea, Robert J. "Human Nature Theory and Political Philosophy." *American Journal of Political Science* 22 (1978): 656–679.

———. *Morality and Human Nature: A New Route to Ethical Theory*. Philadelphia: Temple University Press, 1990.

Mead, Margaret. *Coming of Age in Samoa: A Psychological Study of Primitive Youth for Western Civilization*. New York: William Morrow, 1928.

Mednick, Sarnoff, and William Gabrielli. "Genetic Influences in Criminal Convictions: Evidence from an Adoption Cohort." *Science* 224 (1984): 891–894.

Mednick, Sarnoff, and Terrie E. Moffit. *The Causes of Crime: New Biological Approaches.* New York: Cambridge University Press, 1987.

Melzer, Arthur M., et al., eds. *Technology in the Western Political Tradition.* Ithaca, N.Y.: Cornell University Press, 1993.

Miller, Barbara D. *The Endangered Sex: Neglect of Female Children in Rural Northern India.* Ithaca and London: Cornell University Press, 1981.

Miller, Henry I. "A Need to Reinvent Biotechnology Regulation at the EPA." *Science* 266 (1994): 1815–1819.

———. "A Rational Approach to Labeling Biotech-Derived Foods." *Science* 284 (1999): 1471–1472.

Miller, Henry I., and Gregory Conko. "The Science of Biotechnology Meets the Politics of Global Regulation." *Issues in Science and Technology* 17 (2000): 47–54.

Miller, Michelle D. "The Informed-Consent Policy of the International Conference on Harmonization of Technical Requirements for Registration of Pharmaceuticals for Human Use: Knowledge Is the Best Medicine." *Cornell International Law Journal* 30 (1997): 203–244.

Moore, G. E. *Principia Ethica.* Cambridge: Cambridge University Press, 1903.

Moravec, Hans P. *Robot: Mere Machine to Transcendent Mind.* New York: Oxford University Press, 1999.

Mosher, Steven. *A Mother's Ordeal: One Woman's Fight against China's One-Child Policy.* New York: Harcourt Brace Jovanovich, 1993.

Munro, Neil. "Brain Politics." *National Journal* 33 (2001): 335–339.

Murray, Charles. "Deeper into the Brain." *National Review* 52 (2000): 46–49.

———. "IQ and Economic Success." *Public Interest* 128 (1997): 21–35.

Murray, Charles, and Richard J. Herrnstein. *The Bell Curve: Intelligence and Class Structure in American Life.* New York: Free Press, 1995.

Muthulakshmi, R. *Female Infanticide: Its Causes and Solutions.* New Delhi: Discovery Publishing House, 1997.

National Bioethics Advisory Commission. *Cloning Human Beings.* Rockville, Md.: National Bioethics Advisory Commission, 1997.

———. *Ethical and Policy Issues in Research Involving Human Participants, Final Recommendations.* Rockville, Md.: National Bioethics Advisory Commission, 2001.

Neisser, Ulric, ed. *The Rising Curve: Long-Term Gains in IQ and Related Measures.* Washington, D.C.: American Psychological Association, 1998.

Neisser, Ulric, Gweneth Boodoo, et al. "Intelligence: Knowns and Unknowns." *American Psychologist* 51 (1996): 77–101.

Nelkin, Dorothy, and Emily Marden. "Cloning: A Business without Regulation." *Hofstra Law Review* 27 (1999): 569–578.

Newby, Robert G., and Diane E. Newby. "The Bell Curve: Another Chapter in the Continuing Political Economy of Racism." *American Behavioral Scientist* 39 (1995): 12–25.

Nietzsche, Friedrich. *The Portable Nietzsche*, edited by Walter Kaufmann. New York: Viking, 1968.

Norman, Michael. "Living Too Long." *The New York Times Magazine*, January 14, 1996, pp. 36–38.

Nuffield Council on Bioethics. *Genetically Modified Crops: The Ethical and Social Issues*. London, England: Nuffield Council on Bioethics, 1999.

Orwell, George. *1984*. New York: Knopf, 1999.

Paarlberg, Robert. "The Global Food Fight." *Foreign Affairs* 79 (2000): 24–38.

Panigrahi, Lalita. *British Social Policy and Female Infanticide in India*. New Delhi: Munshiram Manoharlal, 1972.

Park, Chai Bin. "Preference for Sons, Family Size, and Sex Ratio: An Empirical Study in Korea." *Demography* 20 (1983): 333–352.

Paul, Diane B. *Controlling Human Heredity: 1865 to the Present*. Atlantic Highlands, N.J.: Humanities Press, 1995.

———. "Eugenic Anxieties, Social Realities, and Political Choices." *Social Research* 59 (1992): 663–683.

Pearson, Karl. *National Life from the Standpoint of Science*. 2d ed. Cambridge: Cambridge University Press, 1919.

Pearson, Veronica. "Population Policy and Eugenics in China." *British Journal of Psychiatry* 167 (1995): 1–4.

Piers, Maria W. *Infanticide*. New York: W. W. Norton, 1978.

Pinker, Steven. *How the Mind Works*. New York: W. W. Norton, 1997.

———. *The Language Instinct*. New York: HarperCollins, 1994.

Pinker, Steven, and Paul Bloom. "Natural Language and Natural Selection." *Behavioral and Brain Sciences* 13 (1990): 707–784.

Plato, *The Republic*.

Plomin, Robert. "Genetics and General Cognitive Ability." *Nature* 402 (1999): C25–C44.

Pool, Ithiel de Sola. *Technologies of Freedom*. Cambridge, Mass.: Harvard/Belknap, 1983.

Posner, Eric A., and Richard A. Posner. "The Demand for Human Cloning." *Hofstra Law Review* 27 (1999): 579–608.

Postrel, Virginia I. *The Future and Its Enemies: The Growing Conflict over Creativity, Enterprise, and Progress*. New York: Touchstone Books, 1999.

Rappley, Marsha, Patricia B. Mullan, et al. "Diagnosis of Attention-Deficit/ Hyperactivity Disorder and Use of Psychotropic Medication in Very Young Children." *Archives of Pediatrics and Adolescent Medicine* 153 (1999): 1039–1045.

Raustiala, Kal, and David Victor. "Biodiversity since Rio: The Future of the Convention on Biological Diversity." *Environment* 38 (1996): 16–30.

Rawls, John. *A Theory of Justice*. Rev. ed. Cambridge, Mass.: Harvard/Belknap, 1999.

Ridley, Matt. *Genome: The Autobiography of a Species in 23 Chapters*. New York: HarperCollins, 2000.

————. *The Red Queen: Sex and the Evolution of Human Nature*. New York: Macmillan, 1993.

Rifkin, Jeremy. *Algeny: A New Word, a New World*. New York: Viking, 1983.

Rifkin, Jeremy, and Ted Howard. *Who Should Play God?* New York: Dell, 1977.

Robertson, John A. *Children of Choice: Freedom and the New Reproductive Technologies*. Princeton, N.J.: Princeton University Press, 1994.

Rose, Michael R. *Evolutionary Biology of Aging*. New York: Oxford University Press, 1991.

————. "Finding the Fountain of Youth." *Technology Review* 95, no. 7 (October 1992): 64–69.

Rosenberg, Alexander. *Darwinism in Philosophy, Social Science, and Policy*. Cambridge: Cambridge University Press, 2000.

Rosenthal, Stephen J. "The Pioneer Fund: Financier of Fascist Research." *American Behavioral Scientist* 39 (1995): 44–62.

Rosman, Lewis. "Public Participation in International Pesticide Regulation: When the Codex Commission Decides." *Virginia Environmental Law Journal* 12 (1993): 329.

Roush, Wade. "Conflict Marks Crime Conference; Charges of Racism and Eugenics Exploded at a Controversial Meeting." *Science* 269 (1995): 1808–1809.

Rowe, David. "A Place at the Policy Table: Behavior Genetics and Estimates of Family Environmental Effects on IQ." *Intelligence* 24 (1997): 133–159.

Runge, C. Ford, and Benjamin Senauer. "A Removable Feast." *Foreign Affairs* 79 (2000): 39–51.

Ruse, Michael. "Biological Species: Natural Kinds, Individuals, or What?" *British Journal for the Philosophy of Science* 38 (1987): 225–242.

Ruse, Michael, and David L. Hull, eds. *The Philosophy of Biology*. New York: Oxford University Press, 1998.

Ruse, Michael, and Edward O. Wilson. "Moral Philosophy as Applied Science: A Darwinian Approach to the Foundations of Ethics." *Philosophy* 61 (1986): 173–192.

Russo, Eugene. "Reconsidering Asilomar." *The Scientist* 14 (April 3, 2000): 15–21.

Sampson, Robert J., and John H. Laub. *Crime in the Making: Pathways and Turning Points Through Life*. Cambridge, Mass.: Harvard University Press, 1993.

Sandel, Michael J. *Democracy's Discontent: America in Search of a Public Philosophy*. Cambridge, Mass.: Harvard University Press, 1996.

Schlesinger, Arthur M., Jr. *The Cycles of American History*. Boston: Houghton Mifflin, 1986.

Schultz, William F. "Comment on Robin Fox." *The National Interest*, no. 63 (Spring 2001): 124–125.

Searle, John R. *The Mystery of Consciousness*. New York: New York Review Books, 1997.

————. *The Rediscovery of Mind* (Cambridge, Mass.: MIT Press, 1992).

Shapiro, Harold T. "Ethical and Policy Issues of Human Cloning." *Science* 277 (1997): 195–197.

Silver, Lee M. *Remaking Eden: Cloning and Beyond in a Brave New World*. New York: Avon, 1998.

Singer, Peter, and Paola Cavalieri. *The Great Ape Project: Equality Beyond Humanity.* New York: St. Martin's Press, 1995.

Singer, Peter, and Helga Kuhse, eds. *Bioethics: An Anthology.* Oxford: Blackwell, 1999.

Singer, Peter, and Susan Reich. *Animal Liberation.* New York: New York Review of Books Press, 1990.

Sloan, Phillip R., eds. *Controlling Our Desires: Historical, Philosophical, Ethical, and Theological Perspectives on the Human Genome Project.* Notre Dame, Ind.: University of Notre Dame Press, 2000.

Sloterdijk, Peter. "Regeln für den Menschenpark: Ein Antwortschreiben zum Brief über den Humanismus." *Die Zeit,* no. 38, September 16, 1999.

Solingen, Etel. "The Political Economy of Nuclear Restraint." *International Security* 19 (1994): 126–169.

Spearman, Charles. *The Abilities of Man: Their Nature and Their Measurement.* New York: Macmillan, 1927.

Stattin, H., and I. Klackenberg-Larsson. "Early Language and Intelligence Development and Their Relationship to Future Criminal Behavior." *Journal of Abnormal Psychology* 102 (1993): 369–378.

Sternberg, Robert J., and Elena L. Grigorenko, eds. *Intelligence, Heredity, and Environment.* Cambridge: Cambridge University Press, 1997.

Stock, Gregory, and John Campbell, eds. *Engineering the Human Germline: An Exploration of the Science and Ethics of Altering the Genes We Pass to Our Children.* New York: Oxford University Press, 2000.

Strauss, William, and Neil Howe. *The Fourth Turning: An American Prophecy.* New York: Broadway Books, 1997.

Symons, Donald. *The Evolution of Human Sexuality.* Oxford: Oxford University Press, 1979.

Talbot, Margaret. "A Desire to Duplicate." *The New York Times Magazine,* February 4, 2001, pp. 40–68.

Taylor, Charles. *Sources of the Self: The Making of the Modern Identity.* Mass.: Harvard University Press, 1989.

Taylor, Sarah E. "FDA Approval Process Ensures Biotech Safety." *Journal of the American Dietetic Association* 100, no. 10 (2000): 3.

Tribe, Laurence H. "Second Thoughts on Cloning." *The New York Times,* December 5, 1997.

Trivers, Robert. "The Evolution of Reciprocal Altruism." *Quarterly Review of Biology* 46 (1971): 35–56.

———. *Social Evolution.* Menlo Park, Calif.: Benjamin/Cummings, 1985.

Uchtmann, Donald L., and Gerald C. Nelson. "US Regulatory Oversight of Agricultural and Food-Related Biotechnology." *American Behavioral Scientist* 44 (2000): 350–377.

Varma, Jay K. "Eugenics and Immigration Restriction: Lessons for Tomorrow." *Journal of the American Medical Association* 275 (1996): 734.

Venter, J. Craig, et al. "The Sequence of the Genome." *Science* 291 (2001): 1304–1351.

Wade, Nicholas. "Of Smart Mice and Even Smarter Men." *The New York Times,* September 7, 1999, p. F1.

———. "A Pill to Extend Life? Don't Dismiss the Notion Too Quickly." *The New York Times*, September 22, 2000, p. A20.

———. "Searching for Genes to Slow the Hands of Biological Time." *The New York Times*, September 26, 2000, p. D1.

Wallace, Helen, and William Wallace. *Policy-Making in the European Union*. Oxford and New York: Oxford University Press, 2000.

Walser, Bryan L. "Shared Technical Decisionmaking and the Disaggregation of Sovereignty." *Tulane Law Review* 72 (1998): 1597–1697.

Wasserman, David. "Science and Social Harm: Genetic Research into Crime and Violence." *Report from the Institute for Philosophy and Public Policy* 15 (1995): 14–19.

Watson, Rory. "EU Institutions Divided on Therapeutic Cloning." *British Medical Journal* 321 (2000): 658.

Weir, Robert F., Susan C. Lawrence, et al., eds. *Genes, Humans, and Self-Knowledge.* Iowa City: University of Iowa Press, 1994.

Wilke, Tom. *Perilous Knowledge: The Human Genome Project and Its Implications.* Berkeley and Los Angeles: University of California Press, 1993.

Wilmut, Ian, Keith Campbell, and Colin Tudge. *The Second Creation: Dolly and the Age of Biological Control.* New York: Farrar, Straus and Giroux, 2000.

Wilson, David Sloan, and Elliott Sober. "Reviving the Superorganism." *Journal of Theoretical Biology* 136 (1989): 337–356.

Wilson, Edward O. *Consilience: The Unity of Knowledge.* New York: Knopf, 1998.

———. *On Human Nature.* Cambridge, Mass.: Harvard University Press, 1978.

———. "Reply to Fukuyama." *The National Interest*, no. 56 (Spring 1999): 35–37.

Wilson, James Q. *Bureaucracy: What Government Agencies Do and Why They Do It.* New York: Basic Books, 1989.

Wilson, James Q. and Richard J. Herrnstein. *Crime and Human Nature.* New York: Simon and Schuster, 1985.

Wingerson, Lois. *Unnatural Selection: The Promise and the Power of Human Gene Research.* New York: Bantam Books, 1998.

Wolfe, Tom. *Hooking Up.* New York: Farrar, Straus and Giroux, 2000.

———. "Sorry, but Your Soul Just Died." *Forbes ASAP*, December 2, 1996.

Wolfson, Adam. "Politics in a Brave New World." *Public Interest* no. 142 (Winter 2001): 31–43.

World Trade Organization. *Trading into the Future.* 2d ed., rev. Lausanne: World Trade Organization, 1999.

Wrangham, Richard, and Dale Peterson. *Demonic Males: Apes and the Origins of Human Violence.* Boston: Houghton Mifflin, 1996.

Wright, Robert. *Nonzero: The Logic of Human Destiny.* New York: Pantheon, 2000.

Wurtzel, Elizabeth. "Adventures in Ritalin." *The New York Times*, April 1, 2000, p. A15.

———. *Prozac Nation: A Memoir.* New York: Riverhead Books, 1994.

Zito, Julie Magno, Daniel J. Safer, et al. "Trends in the Prescribing of Psychotropic Medications to Preschoolers." *Journal of the American Medical Association* 283 (2000): 1025–1060.